Homework

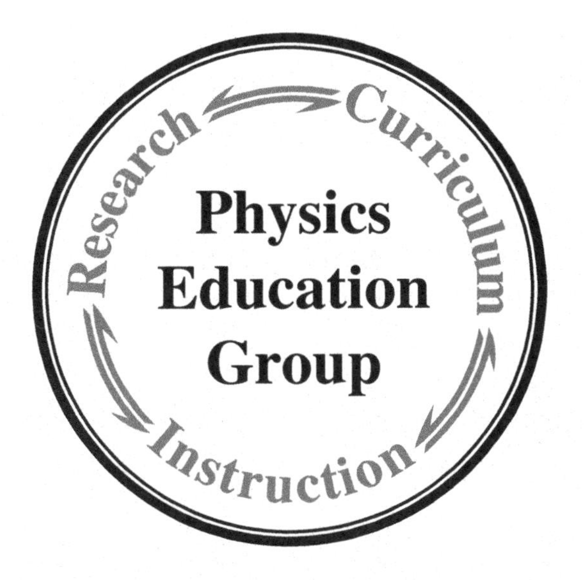

Tutorials in Introductory Physics

First Edition

Lillian C. McDermott, Peter S. Shaffer
and the Physics Education Group

Department of Physics
University of Washington

Prentice Hall
Upper Saddle River, New Jersey 07458

Prentice Hall Series in Educational Innovation

Editor in Chief: *John Challice*
Acquisitions Editor: *Alison Reeves*
Executive Managing Editor: *Kathleen Schiaparelli*
Assistant Managing Editor: *Beth Sturla*
Production Editors: *Shari Toron and Susan Fisher*
Art Director: *Jayne Conte*
Cover Designer: *Bruce Kenselaar*
Manufacturing Manager: *Trudy Pisciotti*
Assistant Manufacturing Manager: *Michael Bell*
Vice President of Production and Manufacturing: *David W. Riccardi*

Upper Saddle River, New Jersey 07458

Printed in the United States of America

22 2022

ISBN:0-13-066245-3

Pearson Education Ltd., *London*
Pearson Education Australia Pty. Limited, *Sydney*
Pearson Education Singapore, Pte. Ltd.
Pearson Education North Asia Ltd., *Hong Kong*
Pearson Education Canada Ltd., *Toronto*
Pearson Educación de Mexico, S.A. de C.V.
Pearson Education—Japan, *Tokyo*
Pearson Education Malaysia, Pte. Ltd.

Preface

Tutorials in Introductory Physics is a set of instructional materials intended to supplement the lectures and textbook of a standard introductory physics course. The emphasis in the tutorials is on the development of important physical concepts and scientific reasoning skills, not on solving the standard quantitative problems found in traditional textbooks.

There is increasing evidence that after instruction in a typical course, many students are unable to apply the physics formalism that they have studied to situations that they have not expressly memorized. In order for meaningful learning to occur, students need more assistance than they can obtain through listening to lectures, reading the textbook, and solving standard quantitative problems. It can be difficult for students who are studying physics for the first time to recognize what they do and do not understand and to learn to ask themselves the types of questions necessary to come to a functional understanding of the material. *Tutorials in Introductory Physics* provides a structure that promotes the active mental engagement of students in the process of learning physics. Questions in the tutorials guide students through the reasoning necessary to construct concepts and to apply them in real-world situations. The tutorials also provide practice in interpreting various representations (*e.g.*, verbal descriptions, diagrams, graphs, and formulas) and in translating back and forth between them. For the most part, the tutorials are intended to be used after concepts have been introduced in the lectures and the laboratory, although most can serve to introduce the topic as well.

The tutorials comprise an integrated system of pretests, worksheets, homework assignments, and post-tests. The tutorial sequence begins with a pretest. These are usually on material already presented in lecture or textbook but not yet covered in tutorial. The pretests help students identify what they do and not understand about the material and what they are expected to learn in the upcoming tutorial. They also inform the instructors about the level of student understanding. The worksheets, which consist of carefully sequenced tasks and questions, provide the structure for the tutorial sessions. Students work together in small groups, constructing answers for themselves through discussions with one another and with the tutorial instructors. The tutorial instructors do not lecture but ask questions designed to help students find their own answers. The tutorial homework reinforces and extends what is covered in the worksheets. For the tutorials to be most effective, it is important that course examinations include questions that emphasize the concepts and reasoning skills developed in the tutorials.

The tutorials are primarily designed for a small class setting but have proved to be adaptable to other instructional environments. The curriculum has been shown to be effective for students in regular and honors sections of calculus-based and algebra-based physics.

The tutorials have been developed through an iterative cycle of: research on the learning and teaching of physics, design of curriculum based on this research, and assessment through rigorous pretesting and post-testing in the classroom. *Tutorials in Introductory Physics* has been developed and tested at the University of Washington and pilot-tested at other colleges and universities.

Comments on the First Edition

Ongoing research has led to modifications to the tutorials and associated homework in the Preliminary Edition of *Tutorials in Introductory Physics*. The First Edition incorporates these changes and also includes several new tutorials on topics covered in the Preliminary Edition. In addition, the First Edition contains a new section with tutorials on topics in hydrostatics, thermal physics, and modern physics.

Acknowledgments

Tutorials in Introductory Physics is the product of close collaboration by many members of the Physics Education Group at the University of Washington. In particular, Paula Heron and Stamatis Vokos, faculty in physics, have played an important role in the development of many tutorials. Significant contributions to the First Edition have also been made by current and former graduate students and post-doctoral research associates: Bradley Ambrose, Andrew Boudreaux, Matt Cochran, Gregory Francis, Stephen Kanim, Christian Kautz, Michael Loverude, Luanna G. Ortiz, Mel Sabella, Rachel Scherr, Mark Somers, John Thompson, and Karen Wosilait. Others, whose work on the Preliminary Edition enriched the First Edition include: Chris Border, Patricia Chastain, Randal Harrington, Pamela Kraus, Graham Oberem, Daryl Pedigo, Tara O'Brien Pride, Christopher Richardson, and Richard Steinberg. Lezlie S. DeWater and Donna Messina, experienced K-12 teachers, have provided many useful insights and suggestions. The assistance of Joan Valles in coordinating the work of the Physics Education Group is deeply appreciated.

The collaboration of other colleagues in the Physics Department has been invaluable. Faculty in the introductory calculus-based sequence, and graduate and undergraduate students who have served as tutorial instructors have made many useful comments. Contributions have also been made by many long-term and short-term visitors to our group. Physics instructors who have pilot-tested the tutorials and have provided valuable feedback over an extended period of time include: John Christopher (University of Kentucky), Romana Crnkovic (Minot State University), William Duxler (Los Angeles Pierce College), Robert Endorf (University of Cincinnati), Gregory Francis (Montana State University), James Freericks and Amy Liu (Georgetown University), Gary Gladding (University of Illinois, Urbana-Champaign), Gregory Kilcup (The Ohio State University), Heidi Mauk (The United States Air Force Academy), Eric Mazur (Harvard University), James Poth (Miami University), and E.F. Redish (University of Maryland).

We thank our editor, Alison Reeves, for her encouragement and advice. We also gratefully acknowledge the support of the National Science Foundation, which has enabled the Physics Education Group to conduct the ongoing, comprehensive program of research, curriculum development, and instruction that has produced *Tutorials in Introductory Physics*. The tutorials have also benefited from the concurrent development of *Physics by Inquiry* (©1996 John Wiley & Sons, Inc.). *Tutorials in Introductory Physics* and *Physics by Inquiry* share a common research base and portions of each have been adapted for the other.

Table of Contents

Part I: Mechanics

Kinematics

Newton's laws

Energy and momentum

Rotation

Part II: Electricity and magnetism

Electrostatics

Electric circuits

Magnetism

Electromagnetism

Part III: Waves

Part IV: Optics

Geometrical optics

Physical optics

Part V: Selected topics

Hydrostatics

Thermodynamics

Modern Physics

Mechanics

1. The position versus time graph at right represents the motion of an object moving in a straight line.

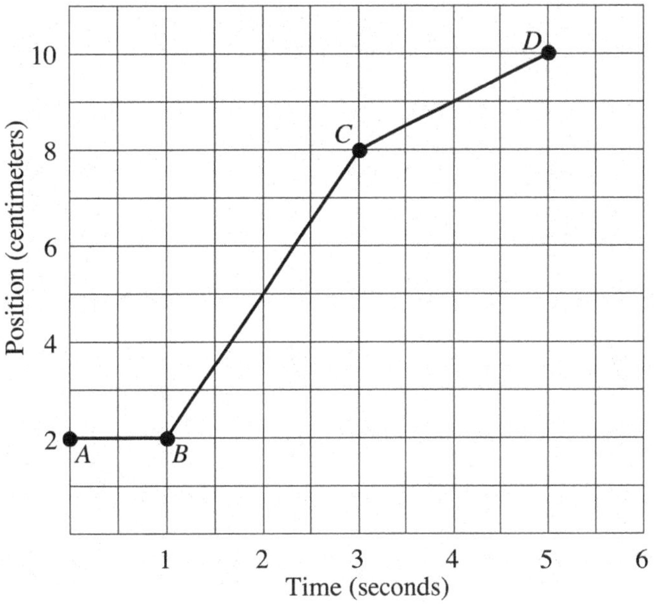

   a. Describe the motion.

   During which periods of time, if any, is the velocity constant? Explain how you can tell.

   Position (centimeters) 10 8 6 4 2 A B C D 1 2 3 4 5 6 Time (seconds)

   b. Find the object's instantaneous velocity at each of the following times. Show your work.

   i. $t = 0.5$ s ii. $t = 2.0$ s iii. $t = 4.0$ s

   How does the method you used to answer parts i–iii rely on your answer to part a?

   c. For each of the following intervals, find the average velocity of the object.

   i. between A and C ii. between A and D iii. between B and D

   On the graph above, sketch and label the lines that would represent an object moving with constant velocity between each of the pairs of points in parts i–iii.

   For each line that you drew, how does the slope compare to the average velocity that you computed above?

   d. In which of the cases from part c, if any, is the average velocity over an interval equal to the average of the constant velocities occurring in that interval? [For example, is $\overline{v}_{AC}$ (the average velocity from A to C) equal to $^1/_2(\overline{v}_{AB} + \overline{v}_{BC})$?]

2. Below is a position versus time graph of the motion of an object that has varying velocity. We will analyze this graph in detail around $t = 2$ s and $x = 2$ cm.

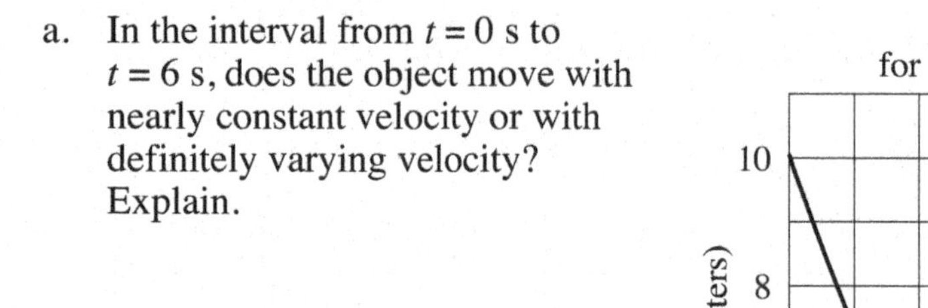

a. In the interval from $t = 0$ s to $t = 6$ s, does the object move with nearly constant velocity or with definitely varying velocity? Explain.

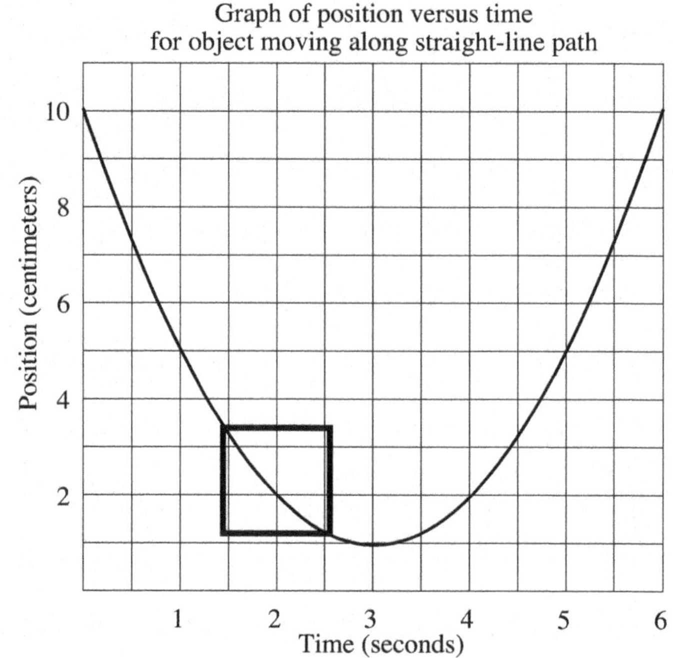

b. In the small box on the graph above is a portion of the graph that corresponds to the motion from $t = 1.5$ s to $t = 2.5$ s.

The position and time coordinates for points in this small interval are given in the following table. Plot these points on the graph below to obtain an expanded view of this small interval.

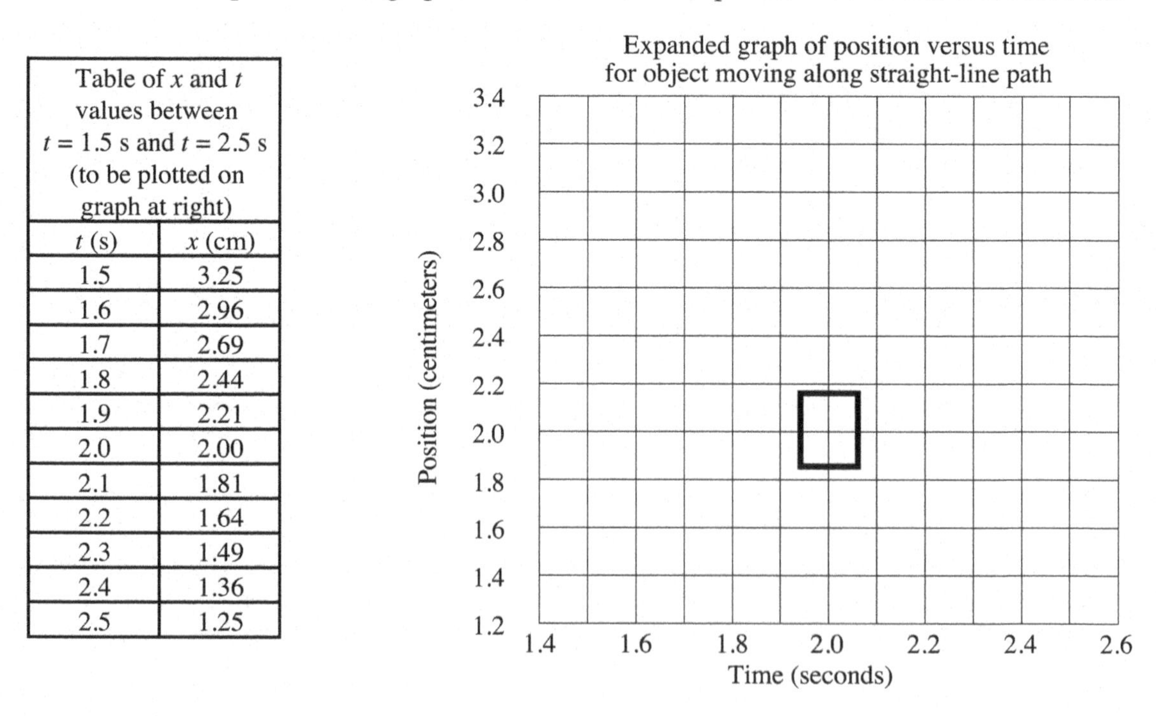

Table of x and t values between $t = 1.5$ s and $t = 2.5$ s (to be plotted on graph at right)

t (s)	x (cm)
1.5	3.25
1.6	2.96
1.7	2.69
1.8	2.44
1.9	2.21
2.0	2.00
2.1	1.81
2.2	1.64
2.3	1.49
2.4	1.36
2.5	1.25

c. Next, we expand the section of the previous graph in the very small box near $t = 2.0$ s.

Position and time coordinates are given below for points in the interval from $t = 1.95$ s to $t = 2.05$ s. Plot these points on the graph provided below.

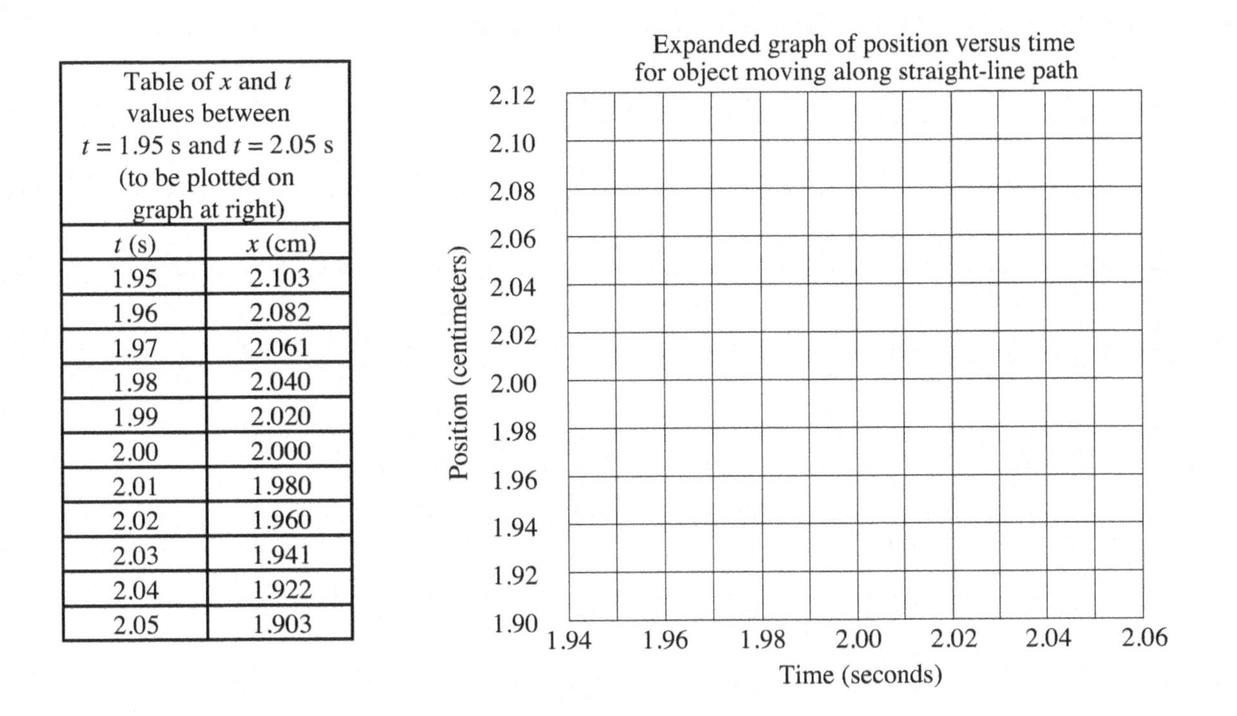

Table of x and t values between $t = 1.95$ s and $t = 2.05$ s (to be plotted on graph at right)	
t (s)	x (cm)
1.95	2.103
1.96	2.082
1.97	2.061
1.98	2.040
1.99	2.020
2.00	2.000
2.01	1.980
2.02	1.960
2.03	1.941
2.04	1.922
2.05	1.903

d. All three graphs are representations of the same motion.

i. How can you account for the last graph being so much straighter than the first?

ii. Can you tell from a very small time interval on a graph whether the motion over the whole graph has constant velocity?

iii. Find the average velocity over the small time interval from $t = 1.95$ s to $t = 2.05$ s. Show your work and explain your reasoning.

How does this average velocity compare to the instantaneous velocity at $t = 2.00$ s? Explain.

3. An object moves along the line from point *1* to point *2* in time Δt.

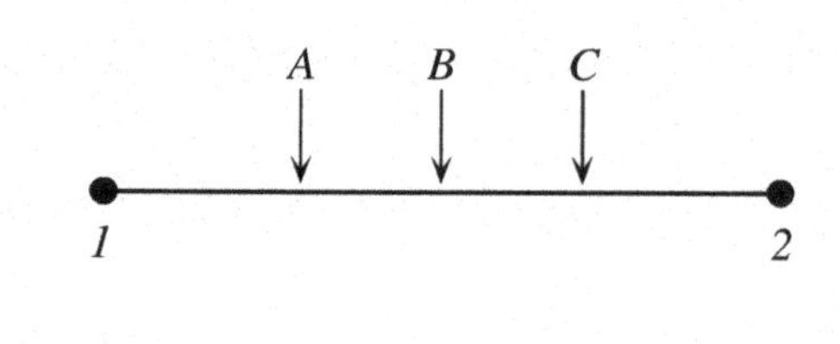

   a. Suppose that the object is *speeding up*. Which of the labeled points *A*, *B*, or *C* *could* correspond to the location of the object at time $\Delta t/2$? (Point *B* lies halfway between points *1* and *2*.) Explain.

   b. Suppose that the object is *slowing down*. Which of the labeled points *A*, *B*, or *C* *could* correspond to the location of the object at time $\Delta t/2$? Explain.

4. Most cars have a speedometer, an odometer, and a clock.

   a. Describe how you could use these devices to determine the instantaneous speed of the car.

   b. Describe how you could use these devices to determine the average speed of the car.

1. In each of the following exercises, a motion will be described in terms of position, velocity, or acceleration. In each case:

   a. Translate the description of the motion into simpler words that describe how you would have to move to produce this motion. If it is not possible to reproduce this motion, explain why not.

   b. Sketch x versus t, v versus t, and a versus t graphs for the motion.

   c. Draw a picture of a track and a ball such that the ball will move with the corresponding motion. Indicate on your diagram:

      - the initial location and initial direction of motion of the ball,
      - the location of $x = 0$, and
      - the positive direction.

The first exercise has been worked as an example.

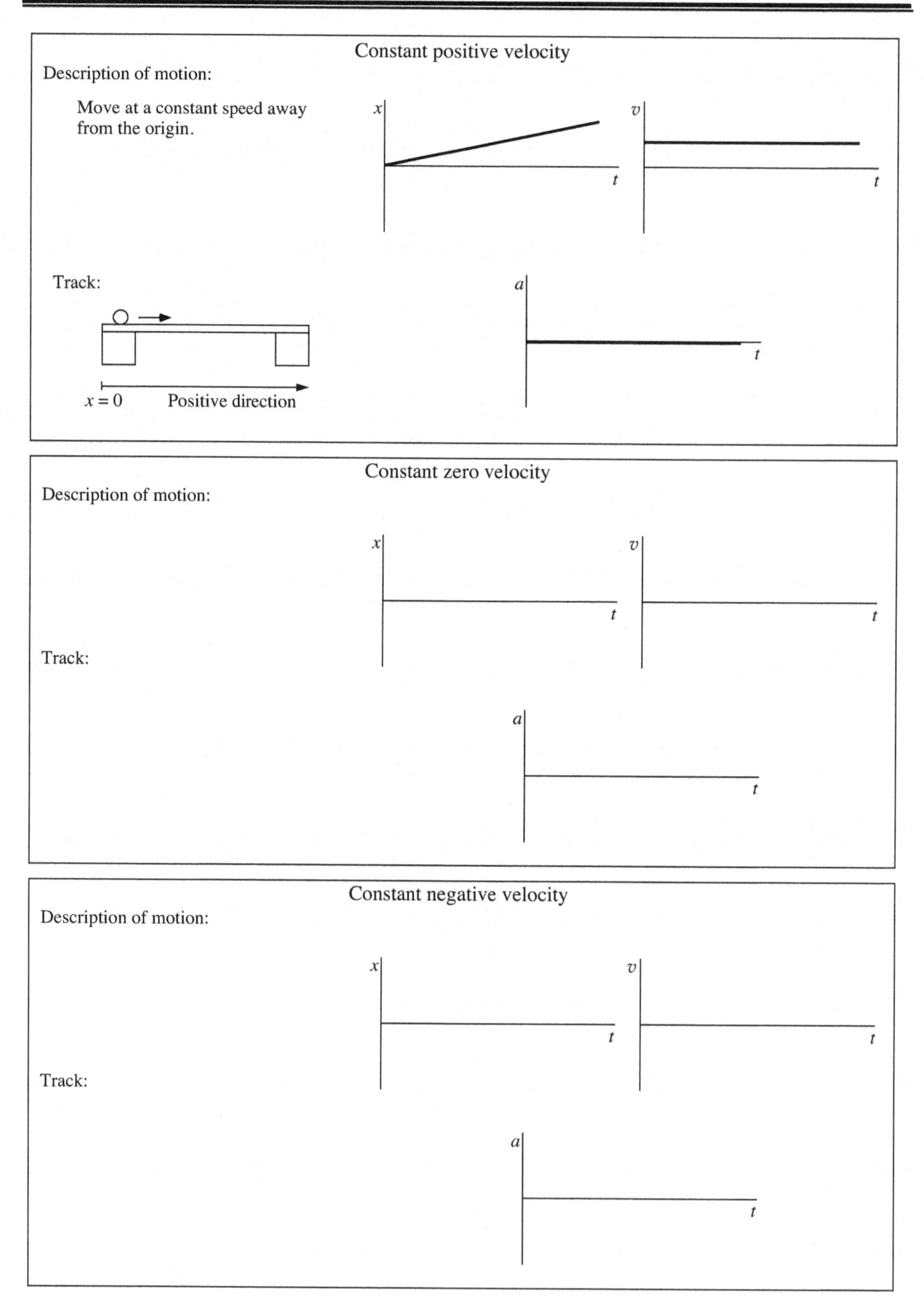
Constant positive velocity
Description of motion:
Move at a constant speed away from the origin.
x
t
v
t
Track:
a
t
x = 0
Positive direction
Constant zero velocity
Description of motion:
x
t
v
t
Track:
a
t
Constant negative velocity
Description of motion:
x
t
v
t
Track:
a
t

Name 

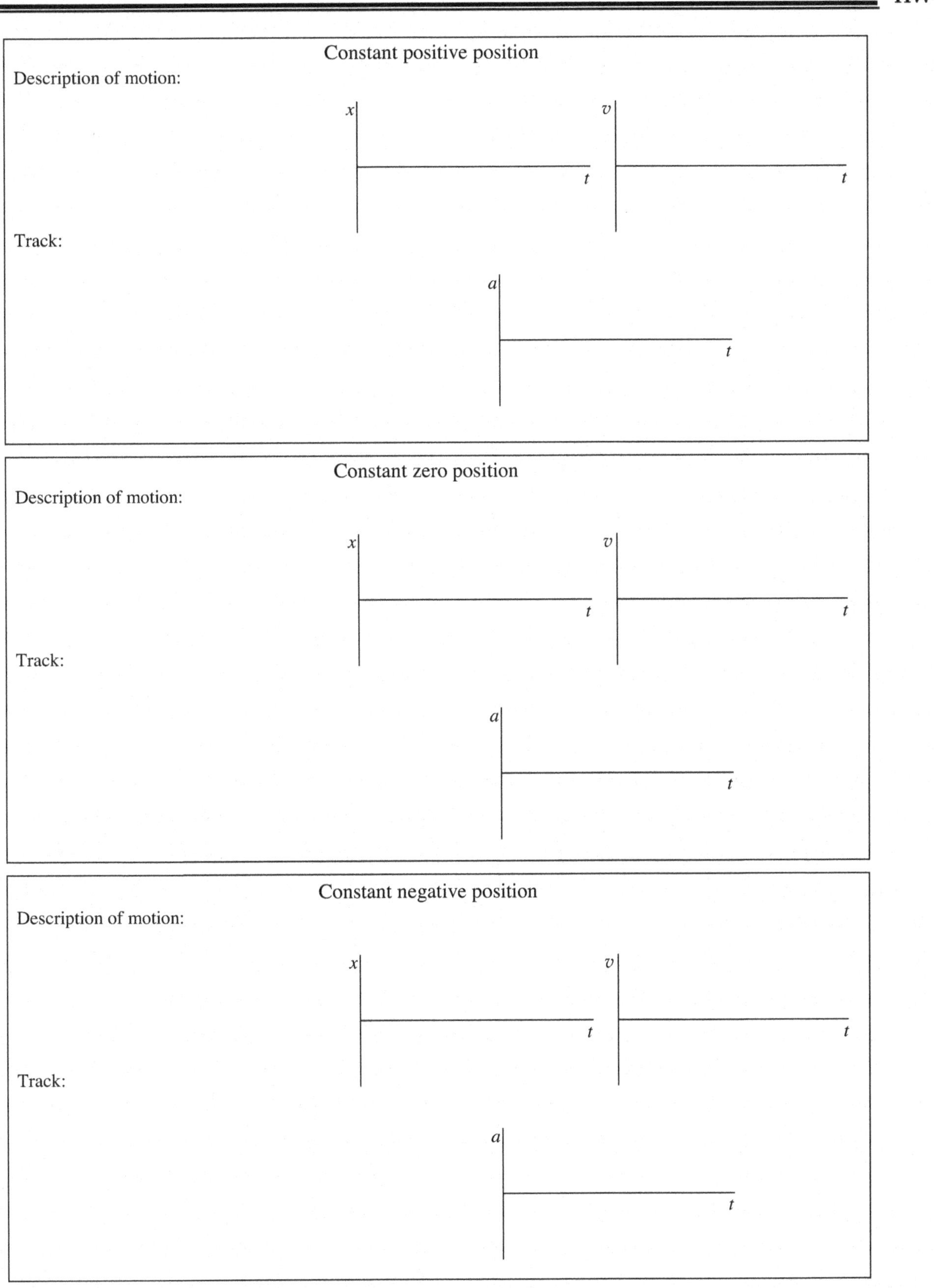

©Prentice Hall, Inc.
First Edition, 2002

Constant positive acceleration

Description of motion:

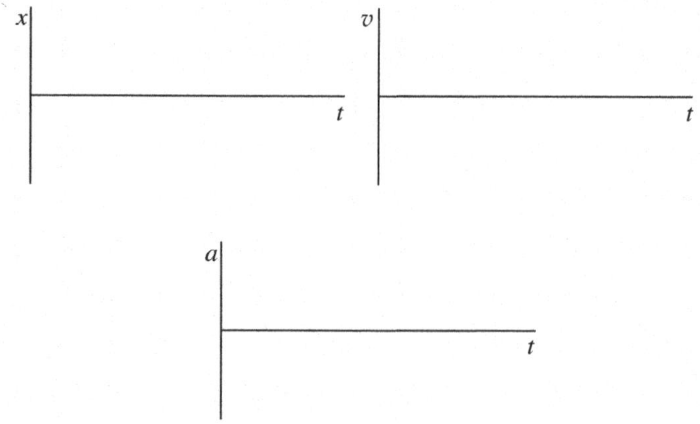

x | t ; v | t

Track:

a | t

Constant zero acceleration

Description of motion:

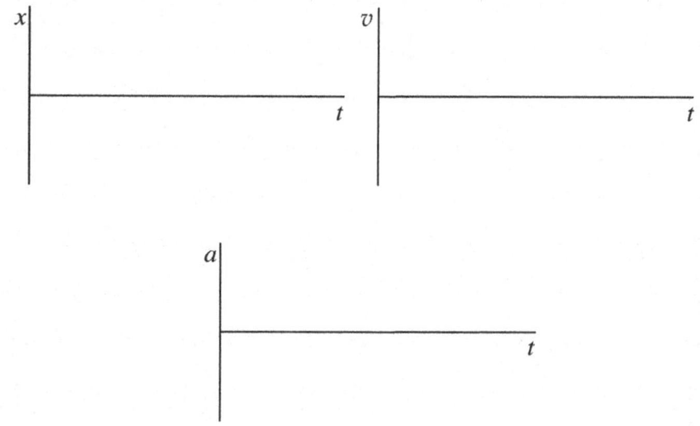

x | t ; v | t

Track:

a | t

Constant negative acceleration

Description of motion:

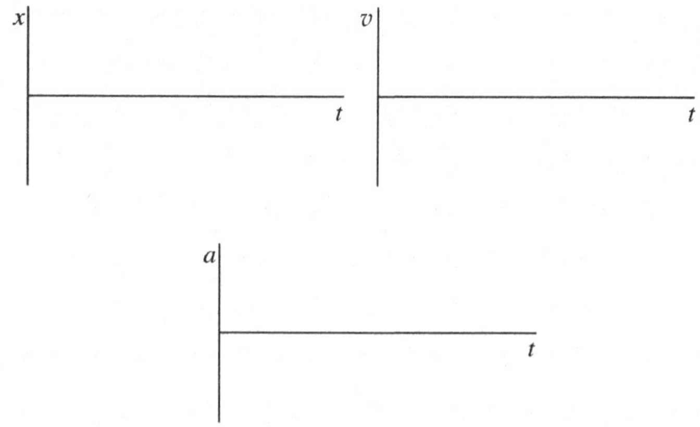

x | t ; v | t

Track:

a | t

2. There are several answers for most of the situations in the previous problem. Find *at least* one other answer to the three motions repeated below.

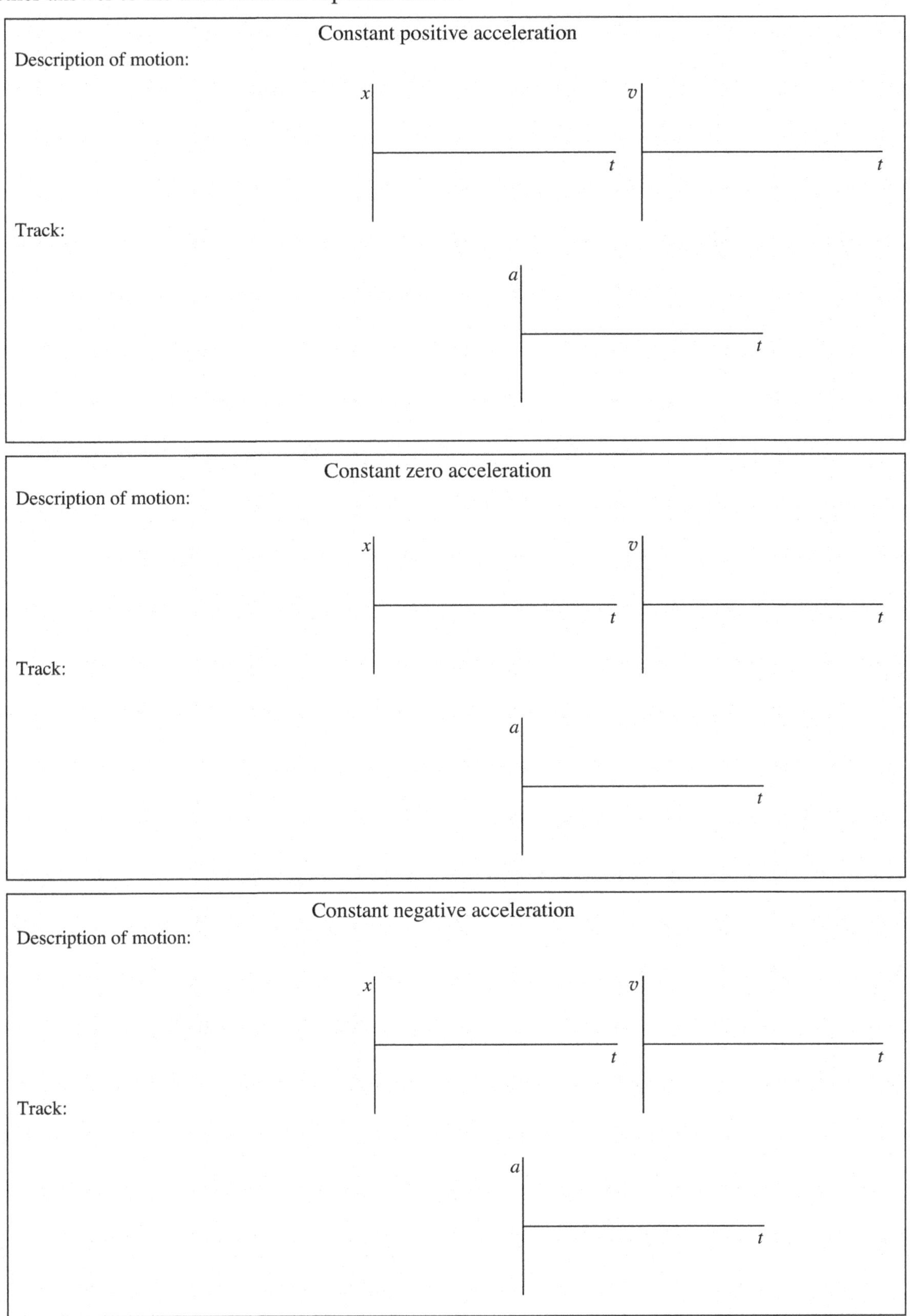

ACCELERATION IN ONE DIMENSION

Name ______________________

1. A ball rolls up, then down an incline. Sketch an *acceleration diagram* for the entire motion. (An *acceleration diagram* is similar to a velocity diagram; however, the vectors on an acceleration diagram represent the *acceleration* rather than the velocity of an object.)

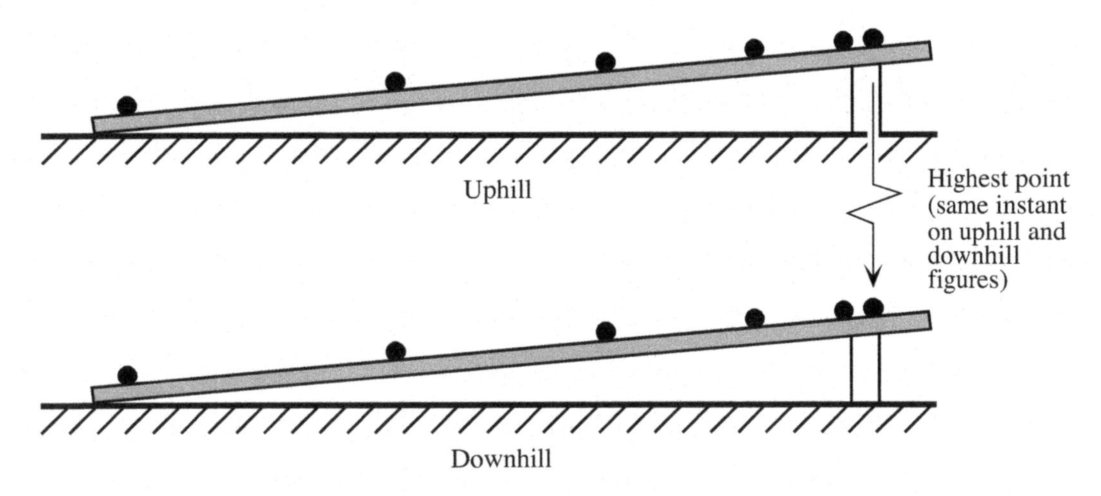

2. Sketch x versus t, v versus t, and a versus t graphs for the entire motion of a ball rolling up and then down an incline.

   a. Use a coordinate system in which the positive x-direction is *down* the track.

   x vs. t v vs. t a vs. t

   b. Use a coordinate system in which the positive x-direction is *up* the track.

   x vs. t v vs. t a vs. t

   c. Can an object have a negative acceleration and be speeding up? If so, describe a possible physical situation and a corresponding coordinate system. If not, explain why not.

3. Describe the motion of an object:

   a. for which the direction of the acceleration is the *same as* the direction of motion of the object.

   b. for which the direction of the acceleration is *opposite to* the direction of motion of the object.

   c. for which the change in velocity is zero.

   d. for which the initial velocity is zero but the acceleration is not zero.

4. Two carts roll toward each other on a level table. The vectors represent the velocities of the carts just before and just after they collide.

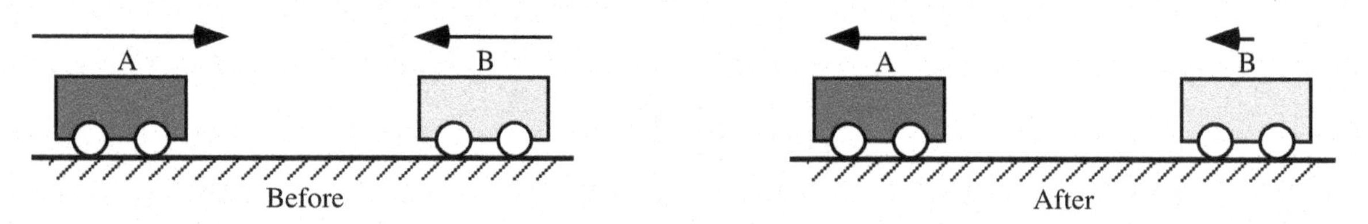

   a. Draw and label a vector for each cart to represent the *change in velocity* from before to after the collision. Make the magnitude and direction of your vectors consistent with the vectors drawn above.

   b. How does the direction of the average acceleration of cart A compare to the direction of the average acceleration of cart B over the time interval shown? Explain.

   c. For the time interval shown, is the magnitude of the average acceleration of cart A *greater than, less than,* or *equal to* the magnitude of the average acceleration of cart B? Explain.

5. In this problem, a cart moves in various ways on a horizontal track. A coordinate system with the positive x-direction to the right is used to measure each motion. For each motion, one of five different representations is given: a strobe diagram, a velocity *versus* time graph, a set of instantaneous velocity vectors, a written description, or a pair of arrows representing the directions of the velocity and acceleration.

Give the remaining *four* representations for each motion. The first exercise has been worked as an example.

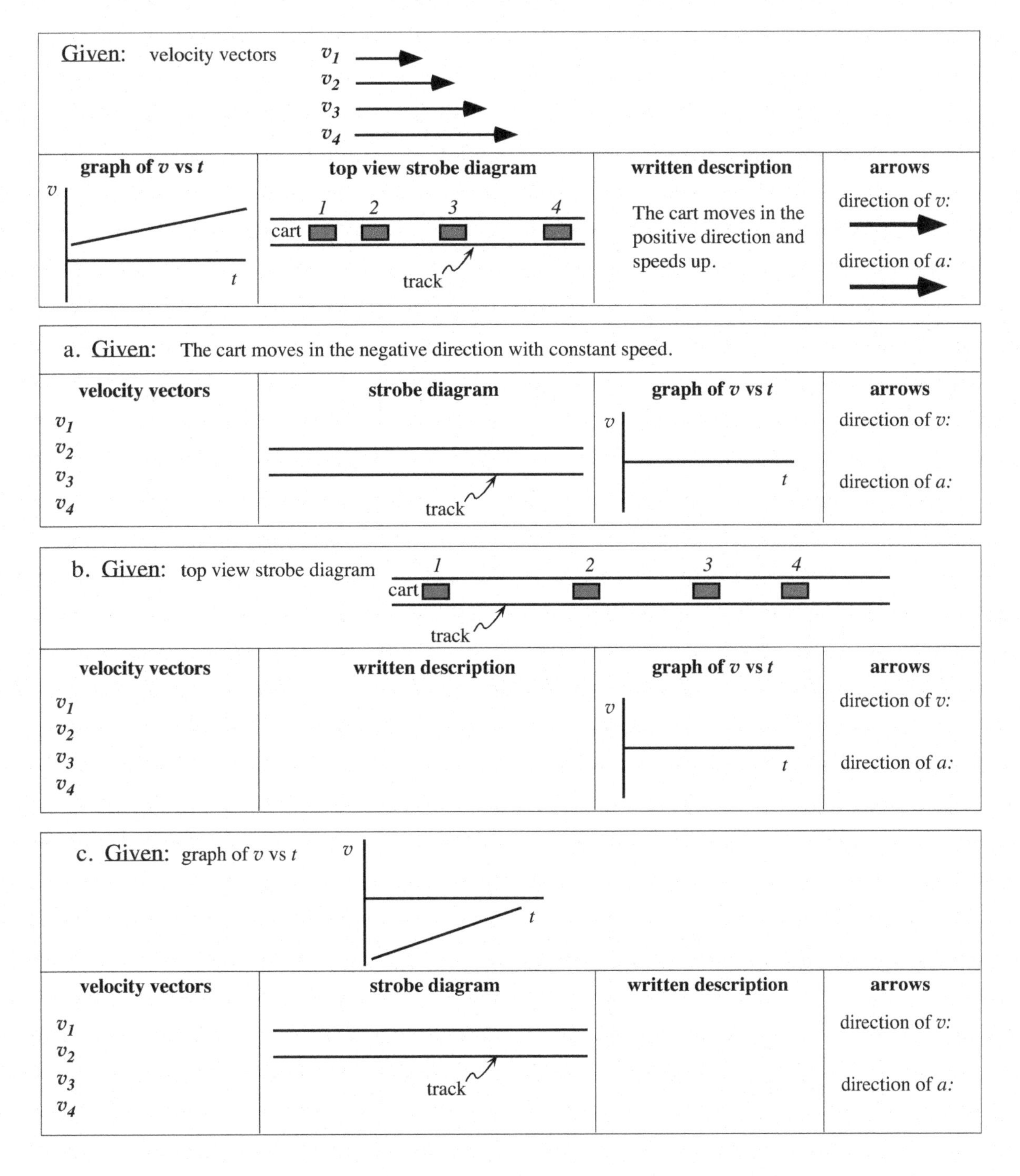

6. Carts A and B move along a horizontal track. The strobe diagram shows the locations of the carts at instants *1–5*, separated by equal time intervals.

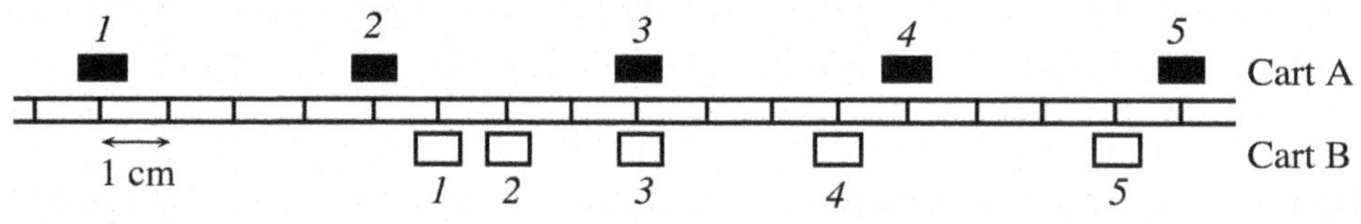

a. At instant *3:*

- is cart A speeding up, slowing down, or moving with constant speed? Explain.

- is cart B speeding up, slowing down, or moving with constant speed? Explain.

b. Is the speed of cart B *greater than, less than,* or *equal to* the speed of cart A:

- at instant *2?* Explain.

- at instant *3?* Explain.

c. During a small time interval from just before instant *2* until just after instant *2,* does the distance between cart A and cart B *increase, decrease,* or *remain the same?* Explain.

Consider the following response to the above question:

> *"For the small interval containing instant 2, cart B is ahead and speeding up, so the distance between the carts must be increasing."*

Do you agree or disagree? Explain.

d. Is there any time interval during which cart A and cart B have the same average velocity? If so, identify the interval(s) and explain. If not, explain why not.

Is there any instant at which cart A and cart B have the same instantaneous velocity? If so, identify the instant(s) (*e.g.*, "at instant *1,*" or "at an instant between *2* and *3*") and explain. If not, explain why not.

7. Two cars, C and D, travel in the same direction on a long, straight section of highway. During a particular time interval Δt_o, car D is ahead of car C and is speeding up while car C is slowing down.

   During the interval Δt_o it is observed that car C *gains* on car D (*i.e.*, the distance between the cars decreases). Explain how this is possible, and give a specific example of such a case.

8. Two cars, P and Q, travel in the same direction on a long, straight section of a highway. Car P passes car Q, and is adjacent to car Q at time t_o.

   a. Suppose that car P and car Q each move with constant speed. At time t_o, is the magnitude of the instantaneous velocity of car P *greater than, less than,* or *equal to* the magnitude of the instantaneous velocity of car Q? Explain.

   b. Suppose instead that car P is moving with constant speed but car Q is speeding up. At time t_o, is the magnitude of the instantaneous velocity of car P *greater than, less than,* or *equal to* the magnitude of the instantaneous velocity of car Q? Explain.

1. An object moves clockwise with *decreasing speed* around an oval track. The velocity vectors at points G and H are shown.

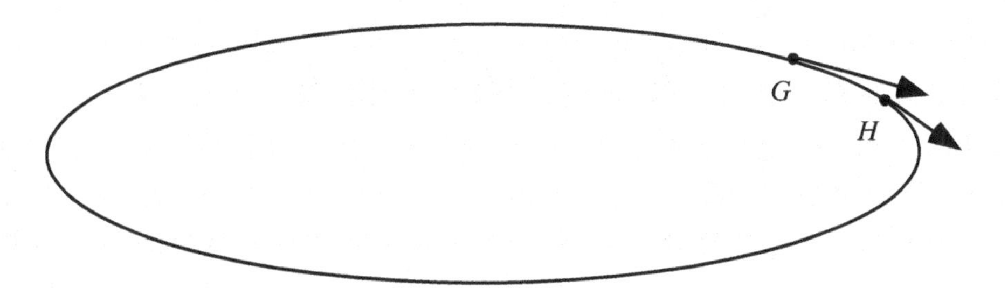

Top view diagram

a. In the space at right, copy the velocity vectors $\vec{v}_G$ and $\vec{v}_H$. From these vectors, determine the change in velocity vector, $\Delta\vec{v}$.

$\vec{v}_G$, $\vec{v}_H$, and $\Delta\vec{v}$

b. If point H were chosen to lie closer to point G, describe how $\Delta\vec{v}$ would change (direction and magnitude).

c. Describe how you would determine the acceleration (direction and magnitude) at point G. In the space at right, indicate the direction of the acceleration of the object at point G.

Direction of acceleration

d. Copy $\vec{v}_G$ and $\vec{v}_H$ (placed "tail-to-tail") in the space at right. How does the angle between the acceleration and velocity vectors compare to 90°? (*i.e.*, Is the angle *greater than, less than,* or *equal to* 90°?)

$\vec{v}_G$ and $\vec{a}_G$
(place them "tail-to-tail")

e. Generalize your results above and from tutorial to answer the following question:

For an object moving along a curved trajectory, how does the angle between the acceleration and velocity vectors compare to 90° if the object moves with (i) constant speed, (ii) increasing speed, and (iii) decreasing speed?

2. Each diagram below shows the velocity and acceleration vectors for an object at a certain instant in time.

	Instant *1*	Instant *2*	Instant *3*	Instant *4*
Acceleration				
Velocity				

a. For each instant, state whether the object is *speeding up, slowing down,* or *moving with constant speed.* Explain your reasoning in each case.

b. The diagram at right illustrates how the acceleration at instant 2 can be treated as having two components—one parallel to the velocity and one perpendicular to the velocity.

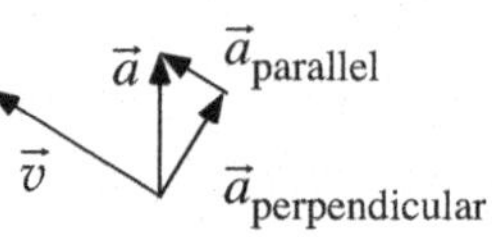

i. For each of the other cases (instants *1*, *3*, and *4*), draw a diagram similar to the one given for instant 2. Label the parallel and perpendicular components of the acceleration relative to the velocity. If either component is zero, state so explicitly.

c. For each of the instants *1–4*, compare your descriptions of the motion in part a with the components of the acceleration that you drew in part b. Then answer the following:

i. Give a general rule for how the component of the acceleration *parallel* to the velocity affects the motion of an object.

ii. Give a general rule for how the component of the acceleration *perpendicular* to the velocity affects the motion of an object.

3. An object starts *from rest* at point *F* and speeds up continuously as it moves around an oval.

$\vec{v}_F = 0$

F

Top view diagram

   a. Choose a point about 1/8th of the way around the oval from point *F*, and label it point *G*. Draw a vector to represent the velocity of the object at point *G*.

   b. Determine the change in velocity vector $\Delta\vec{v}$ between points *F* and *G*.

   c. How would you characterize the direction of $\Delta\vec{v}$ as point *G* moves closer and closer to point *F?*

   d. Each of the following statements in *incorrect*. Discuss the flaws in the reasoning.

      i. "The acceleration at point F is zero. As point G becomes closer and closer to point F, the change in velocity vector becomes smaller and smaller. Eventually, it becomes zero."

      ii. "The acceleration at point F is perpendicular to the curve."

4. A car on a circular track starts from rest at point *A* and moves clockwise with increasing speed. The speed of the car is increasing at a constant rate.

   a. On the diagram at right, draw vectors that represent the *velocity* of the car at each labeled point. Explain.

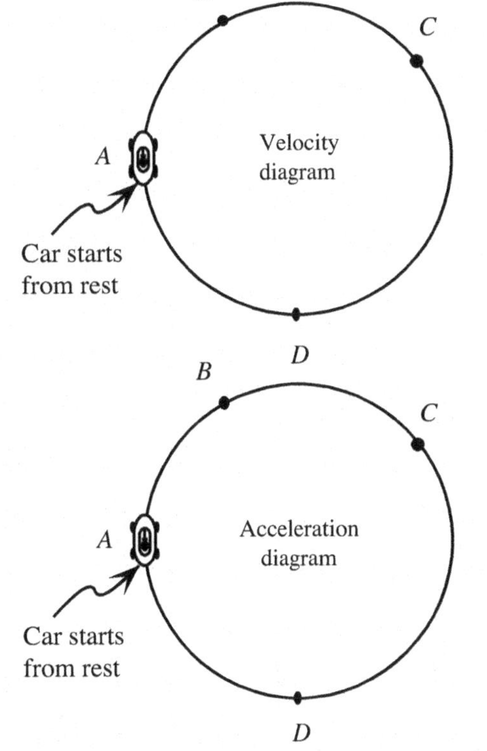

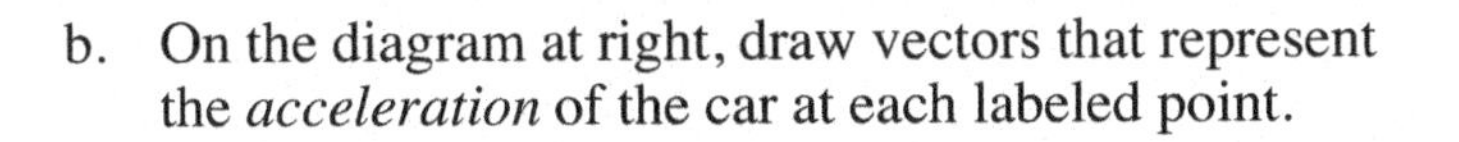

   b. On the diagram at right, draw vectors that represent the *acceleration* of the car at each labeled point.

      i. Explain how your acceleration vector at point *A* is consistent with your answer to question 3 above.

      ii. Explain how the fact that the car is speeding up at a constant rate is represented by the acceleration vectors you have drawn. (*Hint:* Consider the component of the acceleration parallel to the velocity.)

5. An object moves clockwise along the trajectory below (top view shown). The acceleration varies, but is *always* directed toward point *K*.

   a. Draw arrows on the diagram at points *A*–*G* to indicate the *direction* of the acceleration at each point.

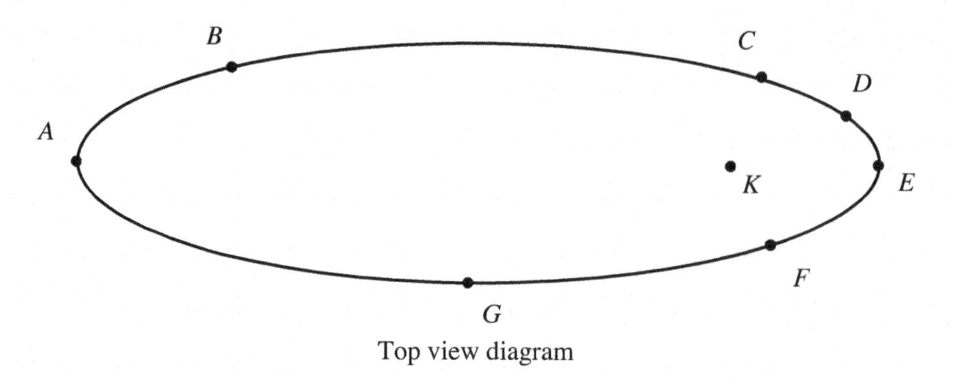

Top view diagram

   b. Next to each of the labeled points, state whether the object is *speeding up, slowing down,* or *moving at constant speed.*

   c. Draw arrows on the diagram below to show the direction of the *velocity* of the object at each labeled point. Draw the arrows with correct relative magnitudes.

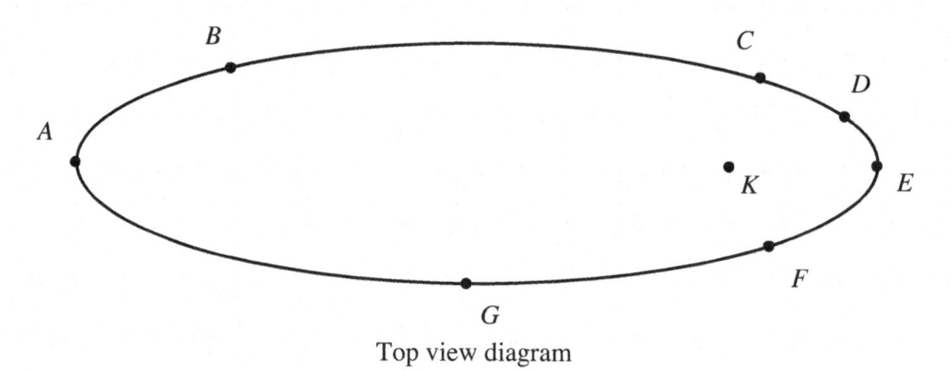

Top view diagram

6. A car travels clockwise once around the track shown below. Starting from rest at point *A,* the car speeds up at a constant rate until just past point *C*. By the time it reaches *D* it is traveling at a constant speed. It then travels at a constant speed the rest of the way around the track.

   a. On the diagram at right, draw velocity vectors for each of the points *A–G*. Be sure that the relative magnitudes of your vectors are consistent.

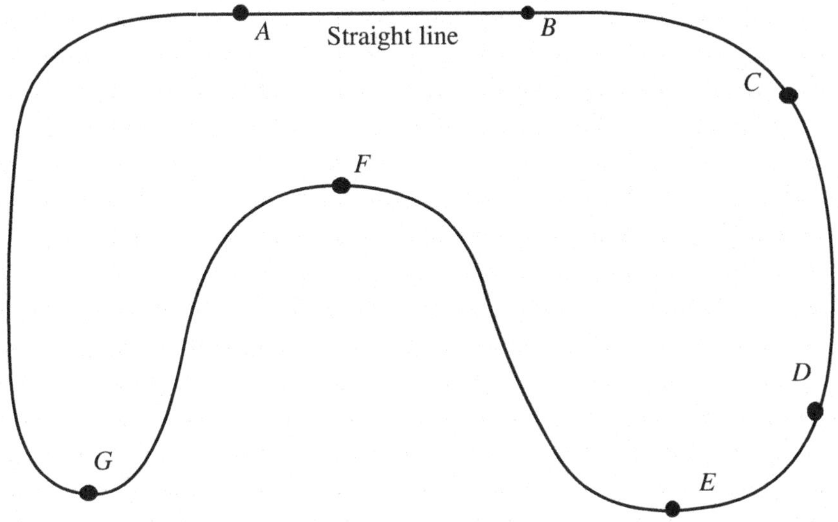

A Straight line B C F D G E

   b. On the diagram at right, draw the acceleration vectors for each of the points *A–G*. If the acceleration is zero at any point(s), indicate that explicitly.

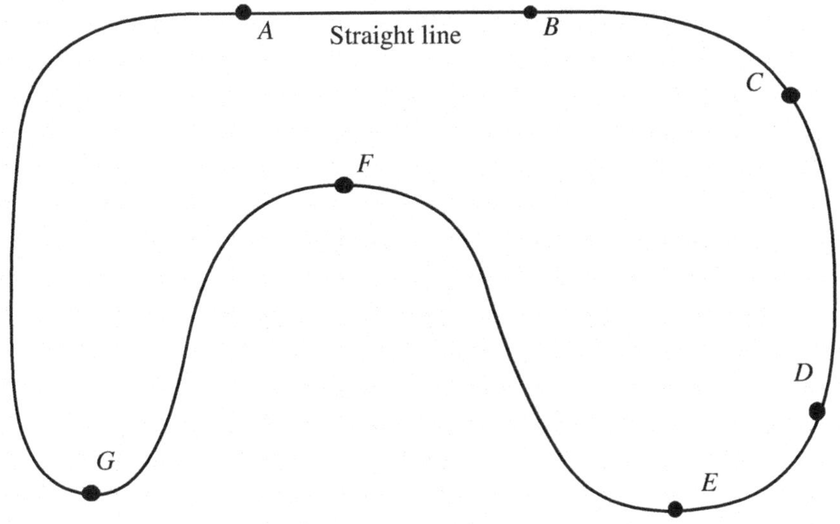

   c. How does the *magnitude* of the acceleration at *E* compare to that at *G?* Explain.

A Straight line B C F D G E

RELATIVE MOTION

Name ____________________ Mech HW–25

1. Two riverboats, A and B, move downstream along a straight section of river as shown. At time t_i, boat A passes a kayak. At time t_f, boat B passes the kayak.

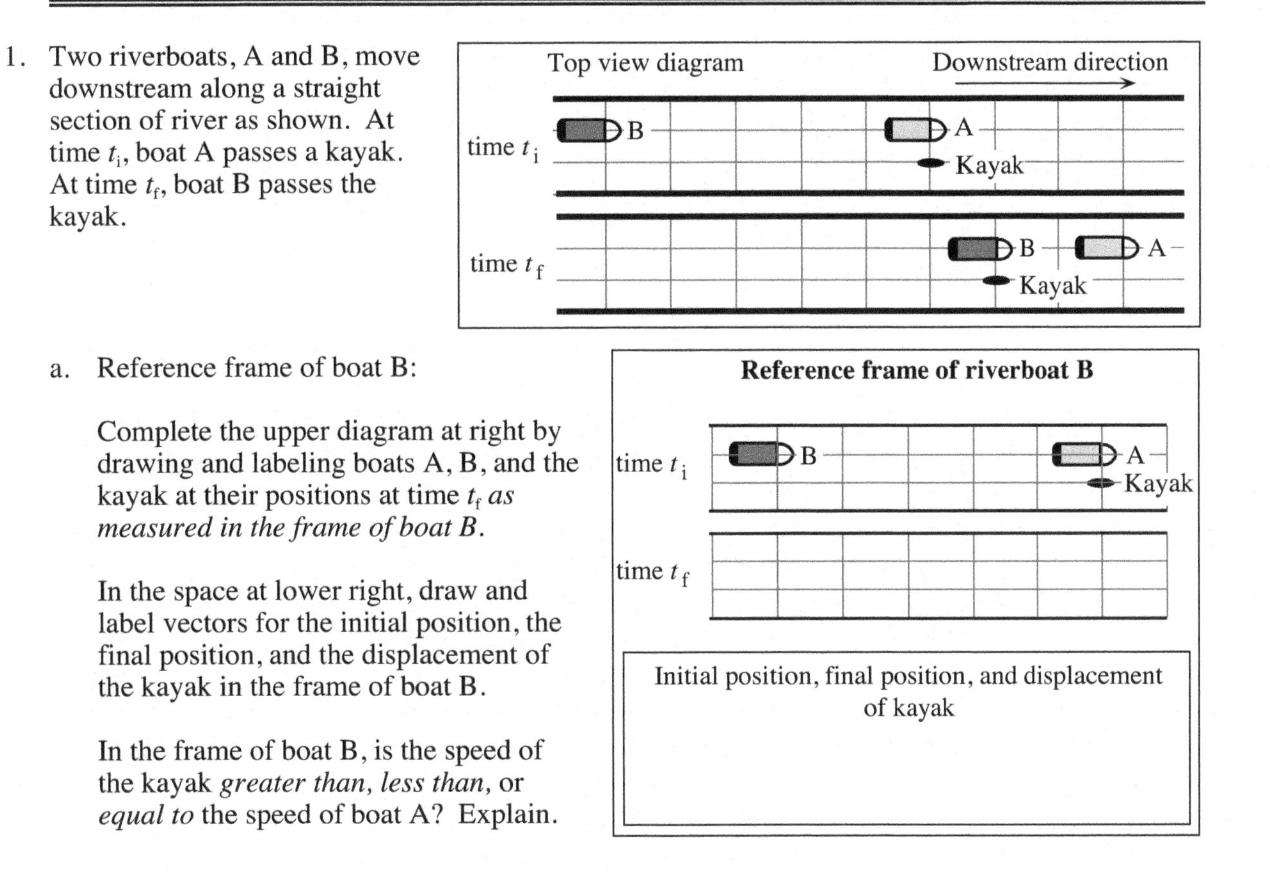

a. Reference frame of boat B:

Complete the upper diagram at right by drawing and labeling boats A, B, and the kayak at their positions at time t_f *as measured in the frame of boat B.*

In the space at lower right, draw and label vectors for the initial position, the final position, and the displacement of the kayak in the frame of boat B.

In the frame of boat B, is the speed of the kayak *greater than, less than,* or *equal to* the speed of boat A? Explain.

b. Reference frame of boat A:

Complete the diagram at right by drawing and labeling the boats and the kayak at their positions at time t_f *as measured in the frame of boat A.*

Reference frame of riverboat A

time t_i B A Kayak

time t_f

c. Is the speed of the kayak in the frame of boat A *greater than, less than,* or *equal to* the speed of the kayak in the frame of boat B? Explain.

d. Rank the following quantities in order of decreasing magnitude from largest to smallest: (i) the displacement of the kayak in the frame of boat A, (ii) the displacement of the kayak in the frame of boat B, (iii) the distance between boat A and boat B at time t_i, and (iv) the distance between boat A and boat B at time t_f. Explain your reasoning.

e. A third riverboat, boat C, moves downstream so as to remain a fixed distance behind boat B at all times. The displacement of the kayak from time t_i to time t_f is measured in the frame of boat A, in the frame of boat B, and in the frame of boat C. Rank these three displacements in order of decreasing magnitude. Explain.

2. A car, a truck, and a traffic cone are on a straight road. Their positions are shown at instants *1–3*, separated by equal time intervals.

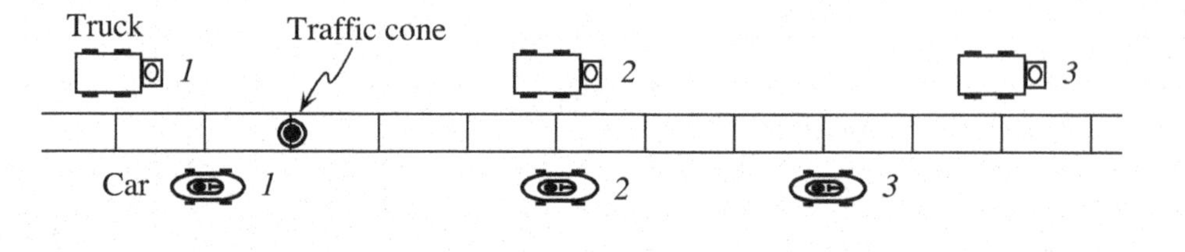

a. In the space provided at right:

• draw the displacement vector of the car *in the frame of the traffic cone* and *in the frame of the truck* for the interval from instant *1* to instant *3*.

• draw the displacement vector of the truck *in the frame of the traffic cone* for the interval from instant *1* to instant *3*.

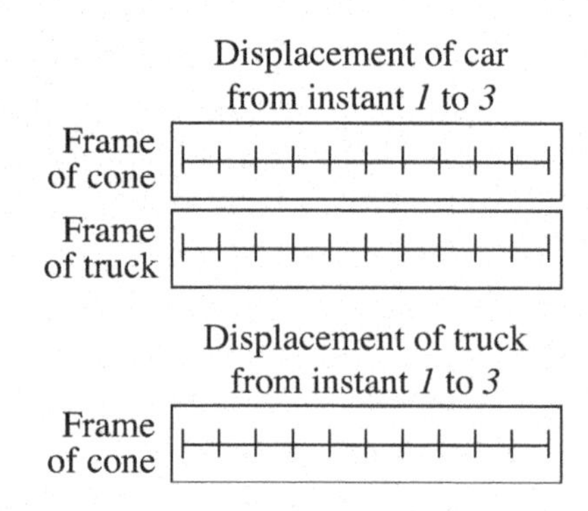

b. In the space below, draw a vector diagram to show which of these three displacement vectors is the sum of the other two.

Express the relationship between the three vectors as an algebraic equation. Use $\Delta\vec{x}_{car,cone}$ to denote the displacement of the car in the frame of the traffic cone, $\Delta\vec{x}_{car,truck}$ to denote the displacement of the car in the frame of the truck, *etc*.

c. The relationship $\vec{v}_{\text{car,cone}} = \vec{v}_{\text{car,truck}} + \vec{v}_{\text{truck,cone}}$ is known as *the Galilean transformation of velocities*. Explain how this relation is consistent with your result above for the displacements.

Does this relationship apply to the *instantaneous* velocities at instant *2?* at instant *3?* Explain.

3. Car P moves to the west with constant speed v_o along a straight road. Car Q starts from rest at instant *1*, and moves to the west with increasing speed. At instant *5*, car Q has speed w_o relative to the road $(w_o < v_o)$. Instants *1–5* are separated by equal time intervals.

At instant *3*, cars P and Q are adjacent to one another *(i.e.*, they have the same position).

a. In the reference frame of the road, at instant *3* is the speed of car Q *greater than, less than,* or *equal to* the speed of car P? Explain.

b. Complete the sketch at right by drawing a qualitatively correct velocity vector for car Q in the frame of the road at instant *3*. Make sure the completed sketch is consistent with your answer to part a.

In this situation, which car is passing the other? Explain.

Sketch of car P and car Q at instant t_3

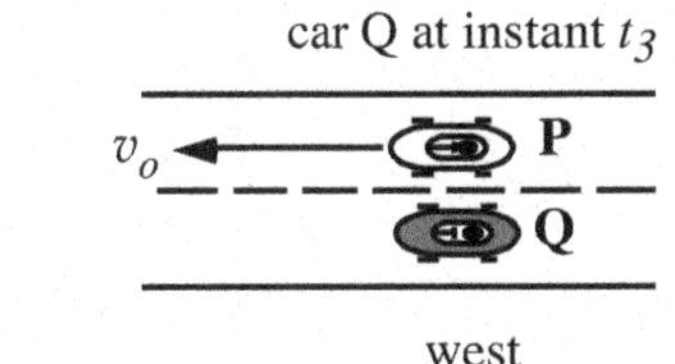

c. In the space at right, sketch and label a vector diagram illustrating the Galilean transformation of velocities that relates $\vec{v}_{P,\text{road}}$, $\vec{v}_{Q,\text{road}}$, and $\vec{v}_{Q,P}$ at instant *3*.

In the frame of car P, at instant *3* is car Q *moving to the west, moving to the east,* or *at rest?* Explain.

Sketch of $\vec{v}_{P,\text{road}}$, $\vec{v}_{Q,\text{road}}$, and $\vec{v}_{Q,P}$ at instant *3*

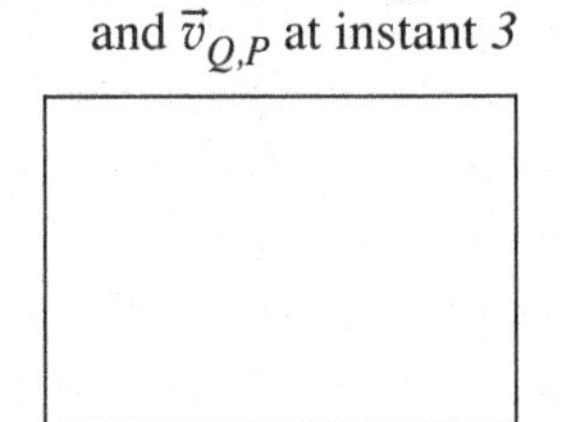

d. Repeat the application of the Galilean transformation to sketch the velocity vectors of car Q in the frame of car P at instants *2*, *3*, and *4*. Explain.

In the frame of car P, is car Q *speeding up, slowing down,* or *moving with constant speed?* Explain.

Velocity of Q in frame of P
at instant *2*
at instant *3*
at instant *4*

e. Complete the diagram at right by drawing car Q at its position at instants *1–5 as measured in the frame of car P*.

Explain how your completed diagram is consistent with your velocity vectors from part d above.

Diagram for the reference frame of car P

Instant *1*	Car P
Instant *2*	
Instant *3*	
Instant *4*	
Instant *5*	

4. A bicycle coasts up a hill while a car drives up the hill at constant speed. The strobe diagram shows their positions at instants *1–4*, separated by equal time intervals. The bicycle comes to rest relative to the road at instant *4*.

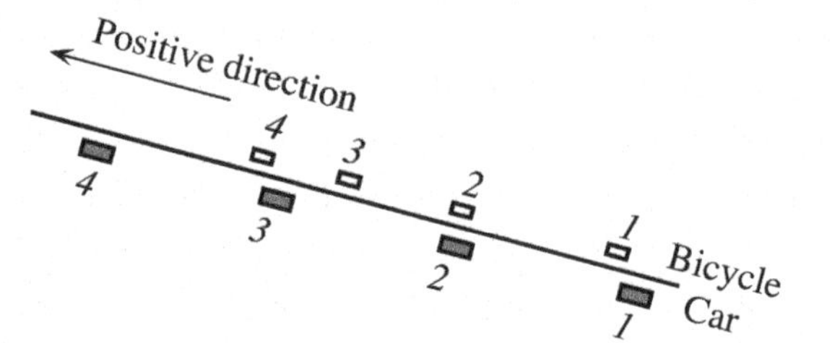

a. As measured in the reference frame of the road:

- is the acceleration of the bicycle *in the positive direction, in the negative direction,* or *zero?* Explain.

- are the velocity and acceleration of the bicycle *in the same* or *in opposite* directions?

b. Sketch velocity vectors of the bicycle *in the reference frame of the car* at instants *2* and *3*. Explain your reasoning.

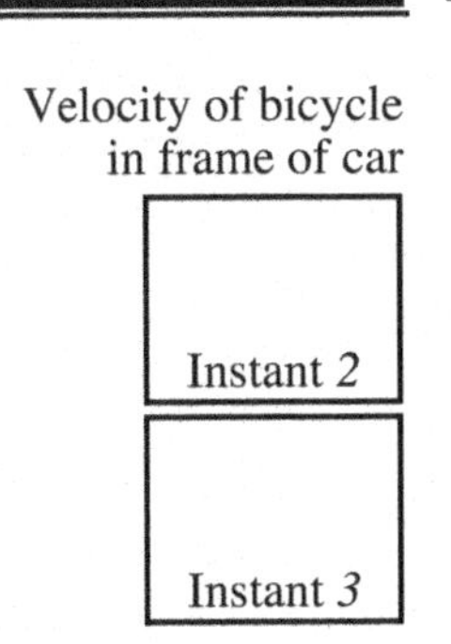

c. In the frame of the car, is the bicycle *moving in the positive direction, moving in the negative direction*, or *at rest:*

- at instant *2?*

- at instant *3?*

d. In the frame of the car, is the bicycle *speeding up, slowing down*, or *moving with constant speed:*

- at instant *2?*

- at instant *3?*

e. In the frame of the car, is the acceleration of the bicycle *in the positive direction, in the negative direction*, or *zero?* Explain how your answer is consistent with your velocity vectors in part b.

In the frame of the car, are the velocity and acceleration of the bicycle *in the same* or *in opposite* directions? Explain how your answer is consistent with your answers to parts c and d.

The frame of the road and the frame of the car are examples of *inertial reference frames*. The direction (and magnitude) of the acceleration of an object is the same in all inertial frames.

f. Consider the following statement:

"The acceleration of the bicycle must be the same in all inertial frames. Since the bicycle is slowing down in the frame of the road, it must be slowing down in the frame of the car as well."

Do you agree or disagree? Explain.

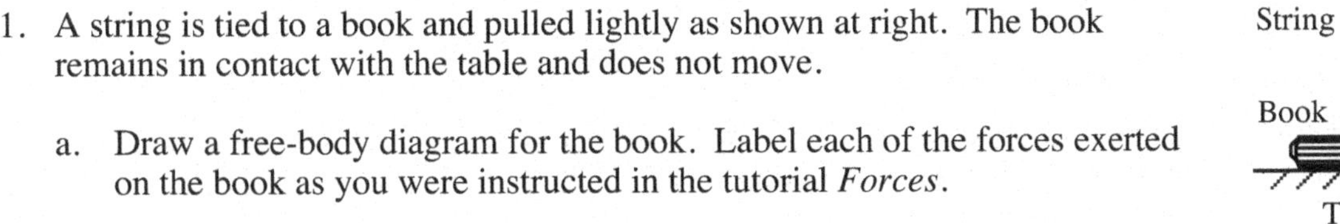

1. A string is tied to a book and pulled lightly as shown at right. The book remains in contact with the table and does not move.

   a. Draw a free-body diagram for the book. Label each of the forces exerted on the book as you were instructed in the tutorial *Forces*.

   b. How do the forces exerted on the book in this case compare to the forces exerted on the book when the string is not present? List any forces that are the same (*i.e.*, same type of force, same direction, and same magnitude) in both cases. Make a separate list of forces that change (or are not present) when the string is pulled.

2. a. Consider the following statement made by a student about a book at rest on a level table:

   > *"The two forces exerted on the book are the normal force directed up and the weight of the book directed down. These are equal and opposite to one another. By Newton's third law they are a third law force pair, so the normal force is always equal to the weight of the book."*

   Do you agree with the student? Explain why you agree or disagree.

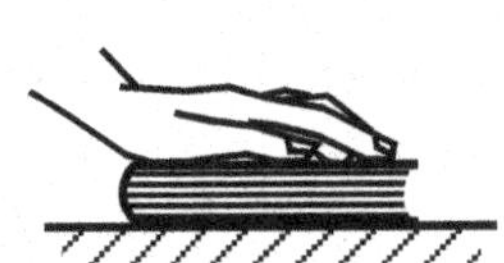

   b. Consider a book on top of a level table while the book is being pressed *straight down* by a hand.

      i. In the space below, draw a free-body diagram for the book. Label the forces as you did in the tutorial *Forces*.

      ii. How do the forces exerted on the book in this case compare to the forces exerted on the book when the hand is not pushing? List any forces that are the same (*i.e.*, same type of force, same direction, and same magnitude) in both cases. Make a separate list of forces that change (or are not present) when the hand is pressing down on the book.

      iii. Is the magnitude of the weight equal to the magnitude of the normal force exerted by the table on the book? How can you tell?

c. Review your answer to part a. In addition, reread the portion of your physics text that discusses Newton's third law. Then consider a book on a level table:

- Which force completes the Newton's third law (or action-reaction) force pair with the normal force exerted on the book by the table?

- Which force completes the Newton's third law (or action-reaction) force pair with the weight of the book?

3. A chain is suspended by a rope as shown at right. The chain is composed of four identical links and does not move.

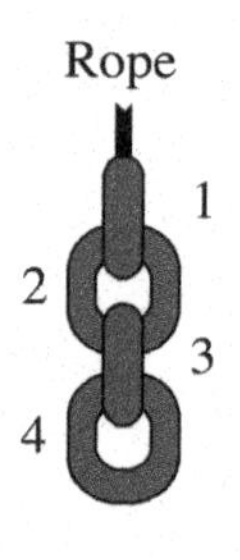

a. In the spaces below, draw a free-body diagram for each of the four links. Label each of the forces as you were instructed in the tutorial *Forces*.

Free-body diagram for link 1	Free-body diagram for link 2	Free-body diagram for link 3	Free-body diagram for link 4

b. Identify all the Newton's third law (action-reaction) force pairs in your diagrams by placing one or more small "X" symbols through each member of the pair (*i.e.*, mark each member of the first pair as —×→, each member of the second pair as —××→, *etc.*).

c. Rank, from largest to smallest, the magnitudes of all the forces on your diagrams. Explain your reasoning, including how you used Newton's second and third laws.

4. Let S_{12} represent the system consisting of links 1 and 2 of the chain in problem 3 (*i.e.*, treat links 1 and 2 as a single object).

Rope

System S_{12}

3

4

a. Draw and label a free-body diagram for system S_{12}.

Free-body diagram for system S_{12}

b. Compare the forces that appear on your free-body diagram for system S_{12} to those that appear on your diagrams for links 1 and 2 in problem 3.

i. For each of the forces that appear on your diagram for system S_{12}, list the corresponding force (or forces) on your diagrams for links 1 and 2.

ii. Are there any forces on your diagrams for links 1 and 2 that you did not list? If so, what characteristic do these forces have in common that none of the others share?

c. Let C represent the system consisting of the whole chain. Draw and label a free-body diagram for C. Make sure that your diagram is consistent with the reasoning you used in part b.

Free-body diagram for system C

5. A block is at rest on an incline as shown below at right. A hand pushes vertically downward with a constant force. The block remains at rest on the incline.

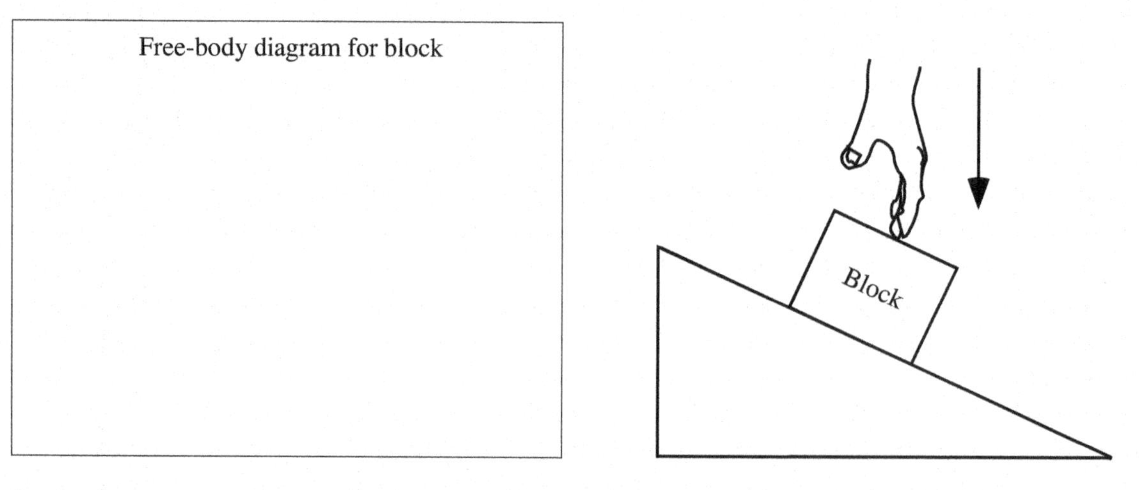

a. In the space provided above, draw a free-body diagram for the block. Label the forces as you did in the tutorial *Forces*.

b. For each force that appears on your free-body diagram, identify the corresponding force that completes the Newton's third law (action-reaction) force pair.

c. Suppose that the hand were to push with a constant force directed as shown at right. The block remains at rest on the incline.

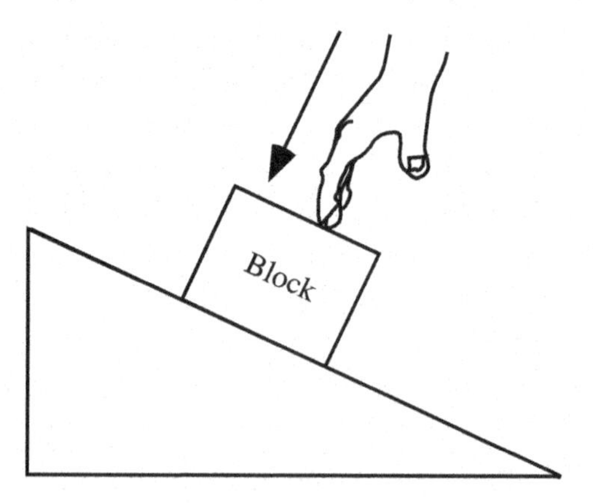

i. Is the magnitude of the net force on the block *greater than, less than,* or *equal to* the magnitude of the net force on the block in part a? Explain.

ii. Is the magnitude of the frictional force exerted on the block by the incline *greater than*, *less than,* or *equal to* the magnitude of the frictional force exerted on the block by the incline in part a? Explain.

6. A person pushes a book against a wall so that the book does not move.

   a. Draw a free-body diagram for the book. Label the forces as you did in the tutorial *Forces*.

   b. For *each* force that appears on your free-body diagram, identify the corresponding force that completes the Newton's third law (or action-reaction) force pair.

1. A block initially at rest is given a quick push by a hand. The block slides across the floor, gradually slows down, and comes to rest.

   a. In the spaces provided, draw and label separate free-body diagrams for the block at each of the three instants shown.

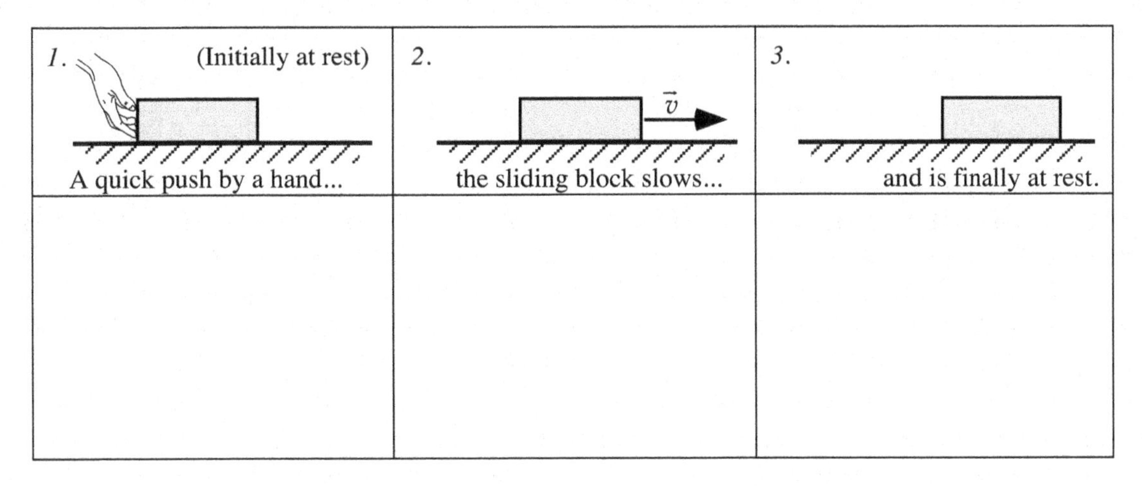

   b. Rank the magnitudes of all the *horizontal* forces in the diagram for instant *1*. Explain.

   c. Are any of the forces that you drew for instant *1* missing from your diagram for instant *2?* If so, for each force that is missing, explain how you knew to include the force on the first diagram but not on the second.

   d. Are any of the forces that you drew for instant *1* missing from your diagram for instant *3?* If so, for each force that is missing, explain how you knew to include the force on the first diagram but not on the third.

2. Two crates, A and B, are in an elevator as shown. The mass of crate A is *greater than* the mass of crate B.

Elevator

A

B

Constant speed

a. The elevator moves downward at *constant speed.*

i. How does the acceleration of crate A compare to that of crate B? Explain.

ii. In the spaces provided below, draw and label separate free-body diagrams for the crates.

Free-body diagram for crate A	Free-body diagram for crate B

iii. Rank the forces on the crates according to magnitude, from largest to smallest. Explain your reasoning, including how you used Newton's second and third laws.

iv. In the spaces provided at right, draw arrows to indicate the direction of the *net force* on each crate. If the net force on either crate is zero, state so explicitly. Explain.

Direction of net force

Crate A	Crate B

Is the magnitude of the *net force* on crate A *greater than, less than,* or *equal to* that on crate B? Explain.

b. As the elevator approaches its destination, its speed decreases. (It continues to move downward.)

i. How does the acceleration of crate A compare to that of crate B? Explain.

ii. In the spaces provided below, draw and label separate free-body diagrams for the crates in this case.

Free-body diagram for crate A	Free-body diagram for crate B

iii. Rank the forces on the crates according to magnitude, from largest to smallest. Explain your reasoning, including how you used Newton's second and third laws.

iv. In the spaces provided at right, draw arrows to indicate the direction of the *net force* on each crate. If the net force on either crate is zero, state so explicitly. Explain.

Direction of net force

Crate A	Crate B

Is the magnitude of the *net force* acting on crate A *greater than, less than,* or *equal to* that on crate B? Explain.

3. A hand pushes three identical bricks as shown. The bricks are moving to the left and speeding up. System A consists of two bricks stacked together. System B consists of a single brick. System C consists of all three bricks. *There is friction between the bricks and the table.*

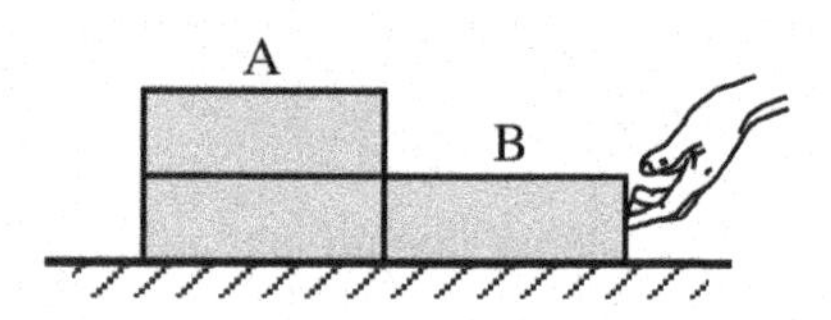

a. In the spaces provided at right, draw and label separate free-body diagrams for systems A and B.

Free-body diagram for system A	Free-body diagram for system B

b. The vector representing the acceleration of system A is shown at right. Draw the acceleration vectors for systems B and C using the same scale. Explain.

Acceleration of A
Acceleration of B
Acceleration of C

c. The vector representing the net force on system A is shown at right. Draw the net force vectors for systems B and C using the same scale. Explain.

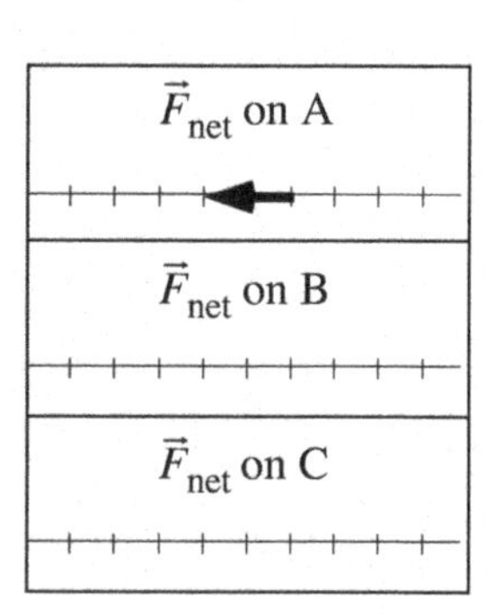

d. The vector representing the frictional force on system A is shown below. Draw the remaining force vectors using the same scale.

$\vec{N}_{BH}$	$\vec{N}_{AB}$	$\vec{N}_{BA}$	$\vec{f}_{AT}$	$\vec{f}_{BT}$

Explain how you knew to draw the force vectors as you did.

4. The table below provides information about the motion of a box in four different situations. In each case, the information given about the motion is in one of the following forms: (1) the algebraic form of Newton's second law, (2) the free-body diagram for the box, or (3) a written description and picture of the physical situation. In each case, complete the table by filling in the information that has been omitted. Case 1 has been done as an example.

(All symbols in the equations represent positive quantities. In each case, use a coordinate system for which +*x* is to the right and +*y* is toward the top of the page.)

KEY: B– box, C– small container, H– hand, S– surface, E– Earth, R, R_1, R_2 – massless ropes

	(1) Algebraic form of Newton's second Law $\vec{F}_{net} = m\vec{a}$	(2) Free-body diagram for box	(3) Written description and picture of physical situation
a. Example	ΣF_x: $F_{BH} - f_{BC} = m_B a_x$ ΣF_y: $N_{BS} - W_{BE} - N_{BC} = 0$	Net force is to the right 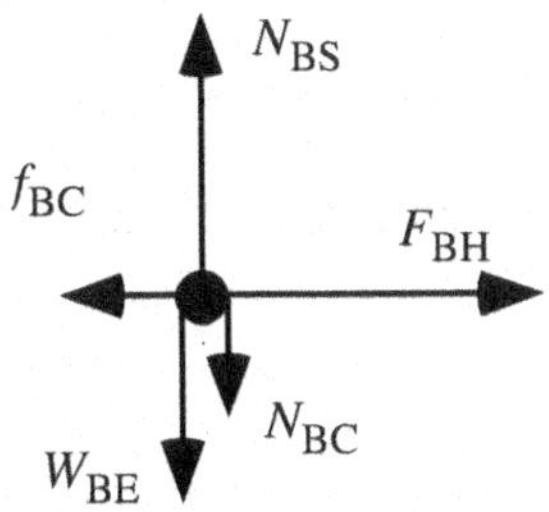 N_{BS}, f_{BC}, F_{BH}, N_{BC}, W_{BE}	A small container is on top of a box. The box is pushed by a hand in the +*x*-direction. There is friction between the container and the box. The box is accelerating to the right on a frictionless surface. 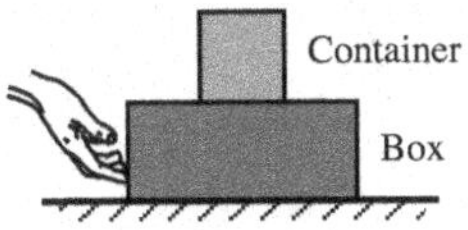 Container Box
b.	ΣF_x: $T_{BR}\cos\theta - F_{BH} = -m_B a_x$ ΣF_y: $T_{BR}\sin\theta + N_{BS} - W_{BE} = 0$	Net force is ________	
c.		Net force is ________	A box is in the back of a truck. The truck accelerates in the +*x*-direction on a straight highway. The box does not move relative to the truck. 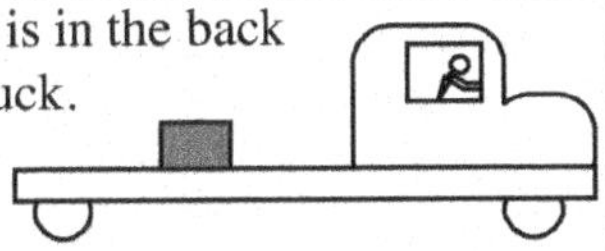
d.		Net force is down 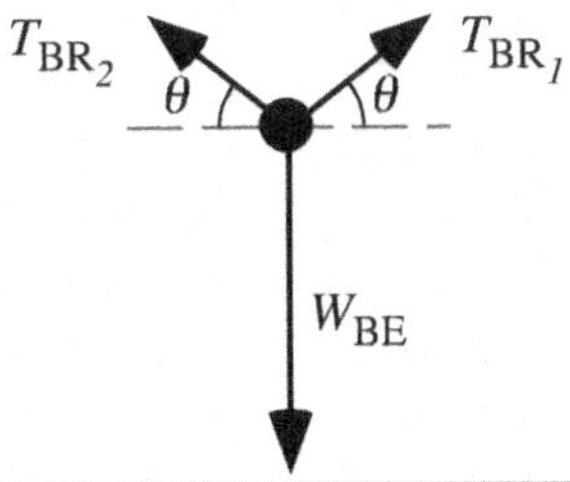 T_{BR_2}, T_{BR_1}, θ, θ, W_{BE}	

5. Two blocks are pushed to the right so that they move together with increasing speed. Block B remains at the height shown. Ignore friction between the ground and block A but not between block A and block B. The mass of block A is 10 kg and the mass of block B is 2 kg. Let system C represent the system consisting of both blocks A and B. (Use $g = 10\ \text{m/s}^2$.)

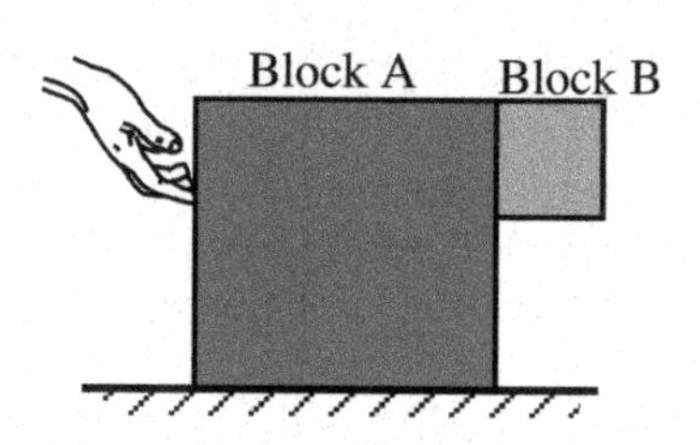

a. For block A, block B, and system C: (1) draw free-body diagrams, (2) identify any Newton's third law force pairs, and (3) write out the algebraic form of Newton's second law.

		Block A	Block B	System C
Free-body diagrams				
Algebraic form of Newton's 2nd law:	*x*:			
	y:			

b. Using *only* the forces in your free-body diagram for system C, calculate the magnitude of the force exerted on system C by the ground (N_{CG}).

c. Using *only* the forces in your free-body diagrams for block A and block B, calculate the magnitude of the force exerted on block A by the ground (N_{AG}).

How should the value of N_{CG} compare to N_{AG}?

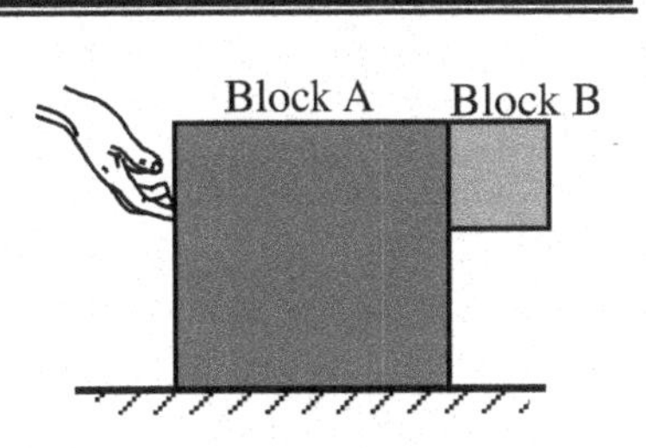

d. Suppose the friction between the two blocks is reduced so that block B slides down as the blocks move to the right. The downward component of the acceleration of block B is 1 m/s^2.

i. For block A and block B: (a) draw new free-body diagrams and (b) write out the algebraic form of Newton's second law.

ii. Is the magnitude of the force exerted on block A by the ground in this case *greater than, less than,* or *equal to* the force exerted on block A by the ground in part c? Explain.

		Block A	Block B
Free-body diagrams			
Algebraic form of Newton's 2nd law:	x:		
	y:		

iii. Calculate the magnitude of the force exerted on block A by the ground. Show your work.

1. A person pulls equally hard on two massless strings that are attached to a block as shown at right. The strings remain taut and symmetrical, forming the same angle with the horizontal at all times.

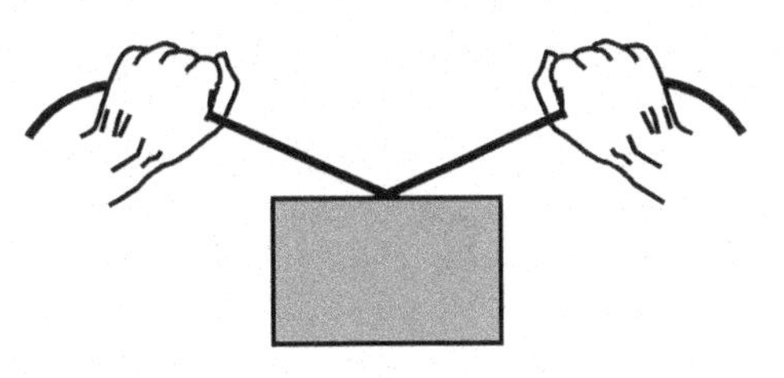

   a. Draw an arrow to indicate the direction of the acceleration of the block for each of the four motions listed below.

Block speeds up while moving downward	Block slows down while moving downward	Block speeds up while moving upward	Block slows down while moving upward

   b. Draw an arrow to indicate the direction of the net force for each of the four motions.

Block speeds up while moving downward	Block slows down while moving downward	Block speeds up while moving upward	Block slows down while moving upward

   Explain the reasoning you used to determine the direction of the net force for each of the four motions.

   c. Draw and label a free-body diagram for the block for each of the four motions. Indicate the relative magnitudes of the forces by the relative lengths of the force vectors. *Draw all diagrams to the same scale.*

Block speeds up while moving downward	Block slows down while moving downward	Block speeds up while moving upward	Block slows down while moving upward

2. Consider the following statement made by a student about each of the four motions in problem 1.

   "The magnitude of the force exerted by the block on an individual string is the same as the magnitude of the force exerted by that string on the block. They are a third law force pair. So I don't understand why the block doesn't move at constant speed if that is true."

   The student has correctly identified a Newton's third law (action-reaction) force pair. Explain why this does not mean that the block must move at constant speed.

3. Five *identical* blocks, each of mass *m*, are pulled across a table as shown. Use the approximations that the table is frictionless and the strings are massless.

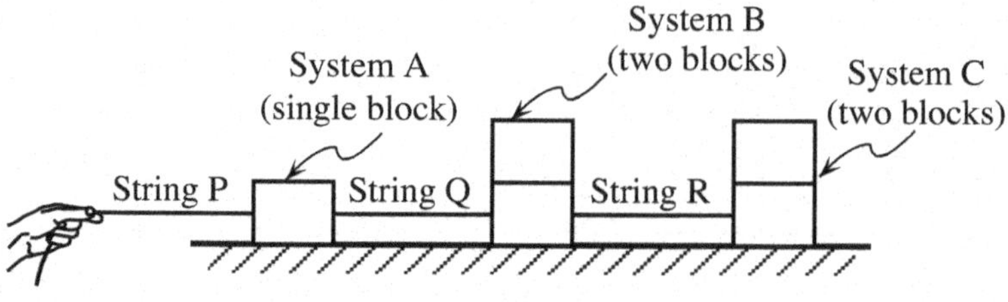

   a. Describe the motion of each of the systems A, B, and C.

   b. Draw vectors below to represent the acceleration of each system.

Acceleration vectors

System A	System B	System C

   c. Draw and label separate free-body diagrams for systems A, B, and C.

Free-body diagram for system A	Free-body diagram for system B	Free-body diagram for system C

   d. Rank the magnitudes of the *net forces* on systems A, B, and C. Explain.

   e. Write expressions for the tension in strings P and R in terms of T_Q, the tension in string Q. Show all your work.

WORK AND THE WORK-ENERGY THEOREM

Name ____________________

1. A hand pushes two blocks, block A and block B, along a frictionless table for a distance d. The mass of block A is greater than the mass of block B ($m_A > m_B$).

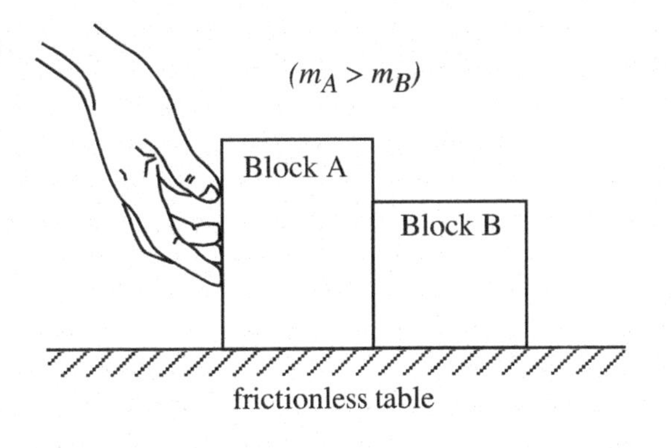

a. Draw a free-body diagram for each block.

Free-body diagram for block A	Free-body diagram for block B

- Does block A do work on block B?
- Does block B do work on block A?

If each block does work on the other, how do the works compare in magnitude and in sign? Explain.

b. How does the net force on block A compare to the net force on block B? Explain.

What does your result suggest about how the *net* work done on block A compares to the *net* work done on block B? Explain.

c. How do the final kinetic energies of the blocks compare? *Base your answer on the work-energy theorem* and your answer to part b above.

d. When the hand starts to push, the blocks are moving with a speed of 2 m/s. Suppose that the work done on block A by the hand during a given displacement is 10 J. Determine the final kinetic energy of each block. Use $m_A = 4$ kg, $m_B = 1$ kg. Show your work.

2. An object moves clockwise along the trajectory shown in the top-view diagram below. The acceleration varies, but is *always* directed toward point K.

   a. Draw and label arrows on the diagram at points A–G to indicate:

   - the direction of the velocity of the object, and
   - the direction of the net force on the object.

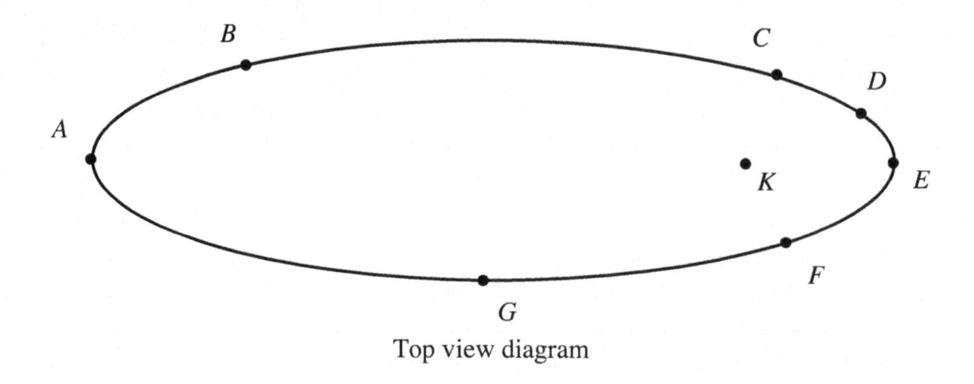

Top view diagram

Explain how you knew to draw the arrows as you did.

   b. For points B, D, and G, determine whether the object is speeding up, slowing down, or moving at constant speed. Explain your reasoning. *Base your answers on the work-energy theorem.*

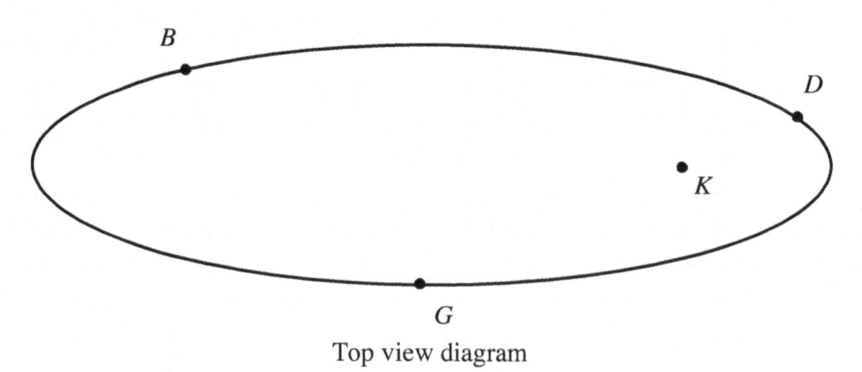

Top view diagram

3. A block is placed on an elevator platform that moves downward with decreasing speed. Consider the time interval Δt_o, in which the speed of the platform changes from v_o to zero.

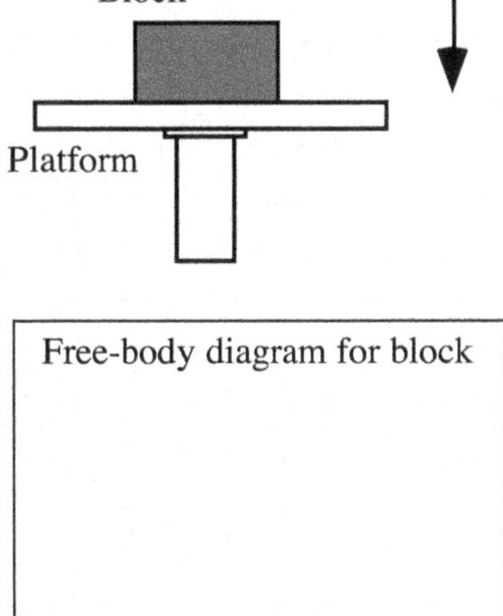

   a. In the space provided, sketch a free-body diagram for the block.

   b. Is the net work done on the block *positive, negative,* or *zero?*

   For each force on the free-body diagram, state whether the work done on the block by that force is *positive, negative,* or *zero*. Explain.

   Free-body diagram for block

   Rank the works you identified above in order of decreasing absolute value. Explain.

   c. Consider reference frame R, moving downward with constant speed v_o.

      i. In the spaces provided, draw arrows to indicate the direction of the velocity and the acceleration of the block in reference frame R during the interval Δt_o. Explain.

In reference frame R:

Direction of *velocity* of block	Direction of *acceleration* of block

      ii. In reference frame R:

   - is the block *speeding up, slowing down,* or *moving with constant speed?* (Base your answer on the directions of the velocity and acceleration).

   - is the change in kinetic energy of the block *positive, negative,* or *zero?* Explain.

      iii. The work-energy theorem can be applied in any inertial frame of reference. Apply the theorem to determine whether the net work done on the block is *positive, negative,* or *zero* in reference frame R. Explain.

iv. In reference frame R, during the interval Δt_o:

- is the displacement (*i.e.*, change in position) of the block *upward, downward,* or *zero?* Explain. (Base your answer on your *velocity* arrow in part c.i.)

- is the net force on the block *upward, downward,* or *zero?* Explain. (Base your answer on your *acceleration* arrow in part c.i.)

- is the work done on the block by the net force *positive, negative,* or *zero?* Explain. (Base your answer on your answers to the previous two questions.)

Make sure your result for the work done on the block by the net force is consistent with your answer to part c.iii.

4. A block is launched up a frictionless ramp, as shown, with initial speed v_o. The block travels up the ramp and continues across the level section.

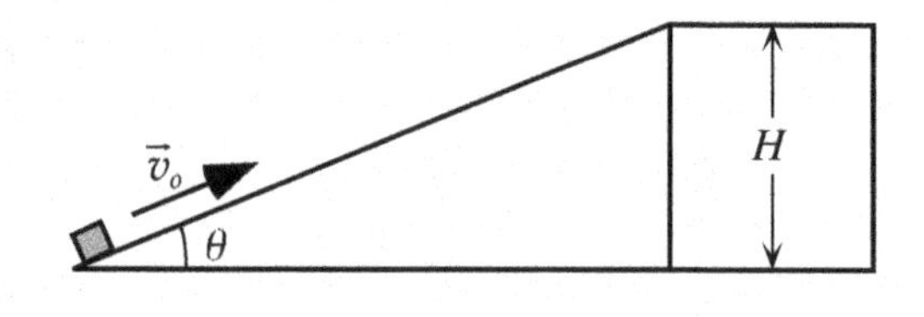

a. List the forces exerted on the block after it has been launched, as it moves up the ramp.

Which forces, if any, do *non-zero* work on the block?

Which forces, if any, do *zero* work on the block?

b. Write an expression for the net work done on the block from the bottom to the top of the ramp. Express your answer in terms of one or more of the following quantities: the weight mg of the block, the angle θ, and the height H of the ramp. Show your work.

5. Suppose the block in the previous problem were launched with the same initial speed on the following frictionless ramps. In each case state whether the *magnitude* of the net work done on the block from the bottom to the top of the ramp is *greater than, less than,* or *equal to* the *magnitude* of the net work done on the block in problem 2. Explain your answer in each case.

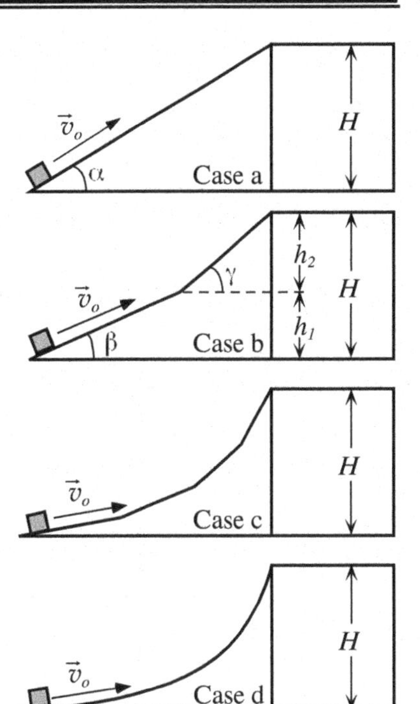

a. The ramp is steeper $(\alpha > \theta)$.

b. The ramp has two sections of different slope.

c. The ramp has several sections of gradually increasing steepness.

d. The ramp is curved.

Use the work-energy theorem to rank the final speeds of the block on the ramps (a–d), assuming the block is launched with the same initial speed in each case. If the final speeds are the same in any of the cases, state that explicitly. Explain.

Name ______________

1. Ball A leaves the edge of a level table with speed v_{A_i} and falls to the floor. At the instant ball A leaves the table edge, another identical ball, B, is released from rest at the height of the table top and also falls to the floor. It is observed that the balls reach the floor at the same time.

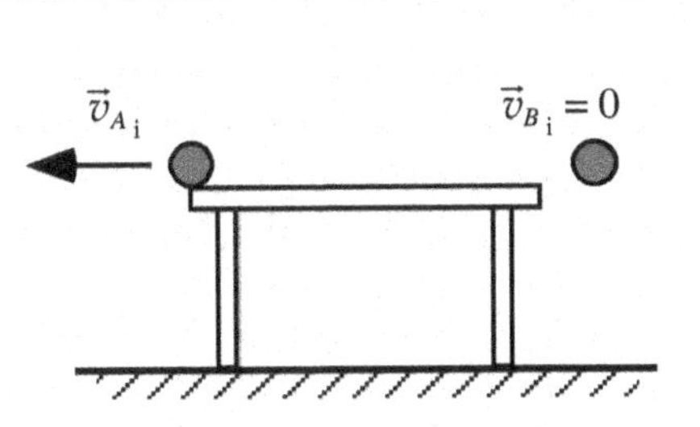

   a. In each question below, consider the interval that *begins* when the balls begin falling and *ends* just before they reach the floor.

      i. Is the magnitude of the impulse imparted to ball B *greater than, less than,* or *equal to* the magnitude of the impulse imparted to ball A? Explain your reasoning.

      ii. In the spaces provided at right, draw an arrow to indicate the direction of the impulse imparted to each ball. Explain your reasoning.

Direction of the impulse imparted to

Ball A	Ball B

      iii. Is the work done on ball B *greater than, less than,* or *equal to* the work done on ball A? Explain your reasoning.

   b. In each question below, consider the balls just before they reach the floor.

      i. Is the magnitude of the momentum of ball B *greater than, less than,* or *equal to* the magnitude of the momentum of ball A? Explain your reasoning.

      ii. Is the kinetic energy of ball B *greater than, less than,* or *equal to* the kinetic energy of ball A? Explain your reasoning.

2. Two identical pucks slide across a level, frictionless table. Initially, the pucks have the same speed, but their velocities are perpendicular to each other as shown.

   The same constant force is exerted on each puck *for the same interval of time:* from the instant shown until *puck 1 crosses the second dotted line*. The pucks remain on the table and do not collide during this time interval.

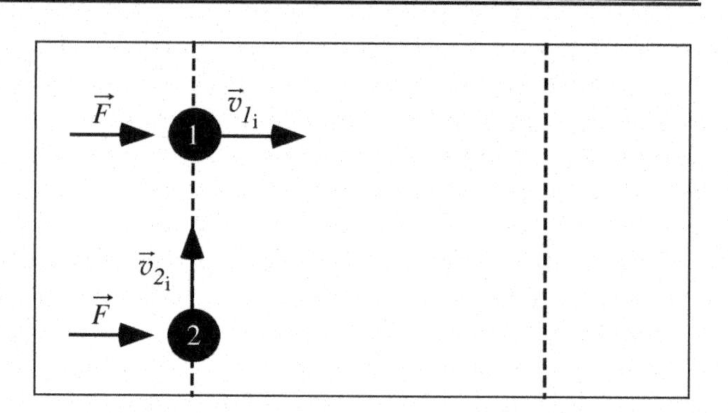

a. When puck 1 crosses the second dotted line, is puck 2 *to the left of, to the right of,* or *crossing* the second dotted line? Explain how you can tell.

   In the diagram above, sketch the trajectory of each of the pucks. Explain how you decided to draw the trajectories the way you did.

b. When puck 1 crosses the second dotted line, is the magnitude of its momentum *greater than, less than,* or *equal to* the magnitude of the momentum of puck 2 at the same instant? Explain the reasoning you used in making this comparison.

c. When puck 1 crosses the second dotted line, is the kinetic energy of puck 1 *greater than, less than,* or *equal to* the kinetic energy of puck 2 at the same instant? Explain the reasoning you used in making this comparison.

CONSERVATION OF MOMENTUM IN ONE DIMENSION

Name ____________

1. Two gliders are on a frictionless, level air track. Initially, glider A moves to the right and glider B is at rest. After the collision, glider A has reversed direction and moves to the left. System C consists of both gliders A and B.

   The mass of glider A is one-fourth the mass of glider B.

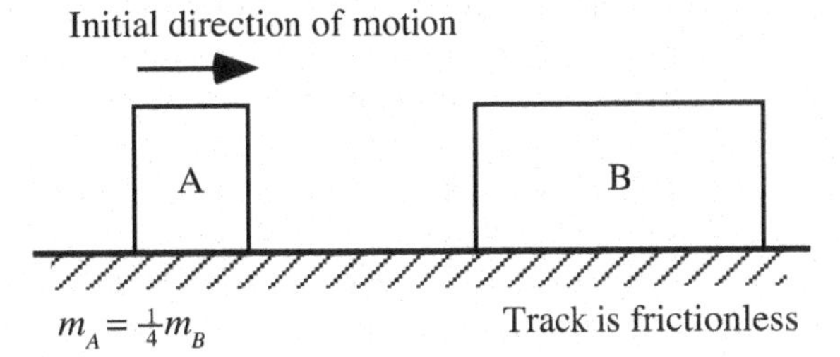

   a. Draw an arrow for each glider and for system C to represent the direction of the *change in momentum* in this collision. Explain how you determined your answer.

Direction of change in momentum

Glider A	Glider B	System C

   Is the magnitude of the change in momentum vector for glider A *greater than, less than,* or *equal to* the magnitude of the change in momentum vector for glider B? Explain.

   b. Draw an arrow for each glider to represent the direction of the *change in velocity* from before to after the collision. Explain how you determined your answer.

Direction of change in velocity

Glider A	Glider B

   Is the magnitude of the change in velocity vector for glider A *greater than, less than,* or *equal to* the magnitude of the change in velocity vector for glider B? Explain.

   c. Consider the following *incorrect* statement:

   > "Glider B will move to the right after this collision, but it would move faster if glider A were to come to a stop, giving glider B all its momentum."

   Describe what is incorrect about this statement and explain how you can tell.

2. a. A firecracker is at rest on a frictionless horizontal table. The firecracker explodes into two pieces of unequal mass that move in opposite directions on the table.

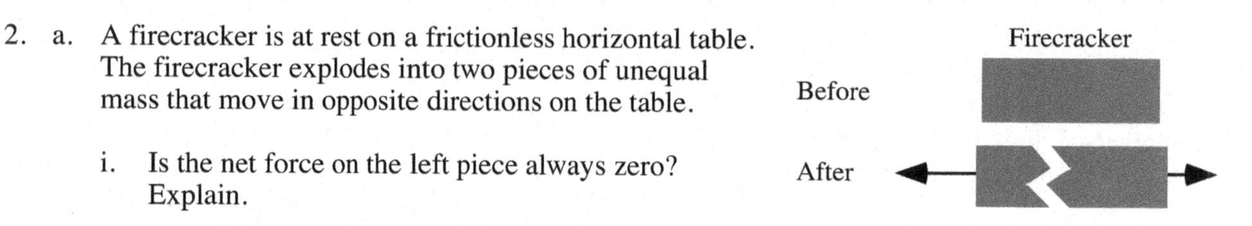

i. Is the net force on the left piece always zero? Explain.

ii. Is the net force on the system consisting of both pieces always zero? Explain.

iii. Is the momentum of the left piece conserved? Explain.

iv. Is the momentum of the system consisting of both pieces conserved? Explain.

b. A block slides down a frictionless incline. The incline is fixed in place on a table.

Incline at rest

i. Is the net force on the block always zero? Explain.

ii. Is the net force on the incline always zero? Explain.

iii. Is the net force on the block-incline system always zero? (*Hint:* Draw free-body diagrams for the block, incline, and system consisting of both objects.) Explain.

iv. Is the momentum of the block conserved? Explain.

v. Is the momentum of the incline conserved? Explain.

vi. Is the momentum of the block-incline system conserved? Explain.

c. Suppose the incline in part b is now placed on a *frictionless* table.

Frictionless

i. Is the net force on the block always zero? Explain.

ii. Is the net force on the incline always zero? Explain.

iii. Is the net force on the block-incline system always zero? Explain.

iv. Is the momentum of the block conserved? Explain.

v. Is the momentum of the incline conserved? Explain.

vi. Is the momentum of the block-incline system conserved? Explain.

d. Two blocks, A and B, are connected by a massless and inextensible string. Their masses are $m_A = 200$ g and $m_B = 400$ g. The blocks are released from rest. The pulley has negligible mass. Let S represent the system of blocks A, B, the string, and the pulley.

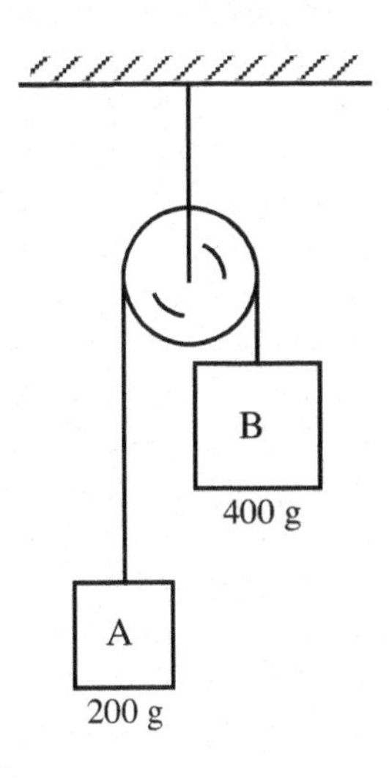

i. Is the net force on block A always zero? Explain.

ii. Is the net force on system S always zero? Explain.

iii. Is the momentum of block A conserved? Explain.

iv. Is the momentum of system S conserved? Explain.

3. Glider A, of mass m, moves to the right with constant speed v_o on a frictionless track toward glider B. Glider B has mass $2m$ and is initially at rest.

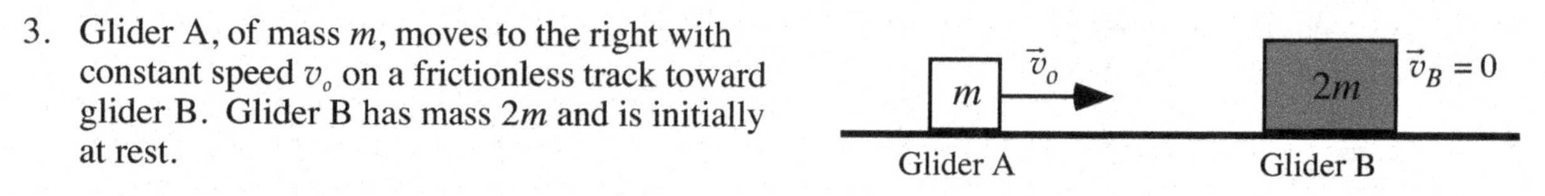

System S consists of gliders A and B.

a. In the spaces provided, draw momentum vectors for glider A, glider B, and system S. Label each vector with its magnitude (express magnitudes in terms of the given quantities m and v_o).

Momentum vectors

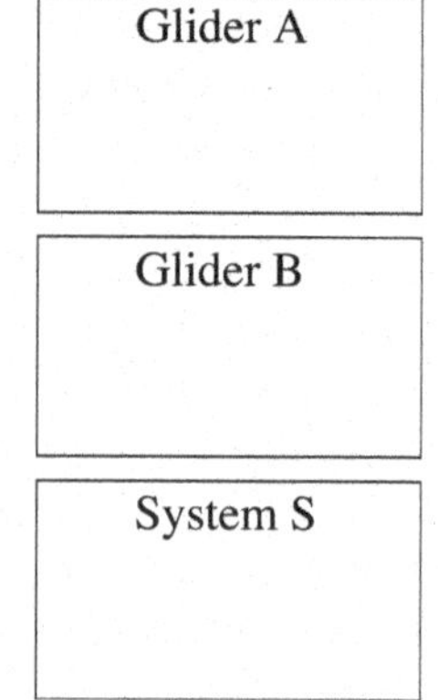

Glider X, of mass $5m$, (not shown in the diagram) moves to the right with speed v_o (*i.e.*, the same speed as glider A) on a second frictionless track parallel to the original track.

b. Apply the Galilean transformation of velocities to determine the velocity vectors of gliders A and B *in the reference frame of glider X*. Draw the vectors in the space at right. Label each vector with its magnitude. (Express the magnitudes in terms of the given quantities.)

Velocity vectors *in the frame of glider X*

Glider A	Glider B

c. Draw *momentum* vectors of gliders A and B in the reference frame of glider X. Label each vector with its magnitude. Explain your reasoning.

Momentum vectors *in the frame of glider X*

Glider A	Glider B

d. Consider the following *incorrect* statement:

"Glider X has momentum $5mv_o$ to the right, so in the reference frame of glider X, the momentum of glider A is $mv_o - 5mv_o = -4mv_o$, or $4mv_o$ to the left."

Explain the error(s) in the reasoning.

Suppose glider X had a different mass (*i.e.*, something other than $5m$). Would the magnitude of the momentum of glider A in the reference frame of glider X be *the same as* or *different than* the value you determined in part c? Explain.

e. Use your momentum vectors from part c to determine the magnitude and direction of the momentum of system S *in the reference frame of glider X*. Explain.

f. Compare your results from part a and part e to answer the following:

- Is the *magnitude* of the momentum of system S the same in the reference frame of glider X as it is in the reference frame of the track? Explain.

- Is the *direction* of the momentum of system S the same in the reference frame of glider X as in the reference frame of the track? Explain.

4. Gliders C and D, of mass 2 kg and 4 kg, respectively, collide on a frictionless track. Glider C initially moves to the right with speed 0.6 m/s relative to the track, while glider D is at rest. After the collision, glider C moves to the left with speed 0.2 m/s. System S consists of gliders C and D.

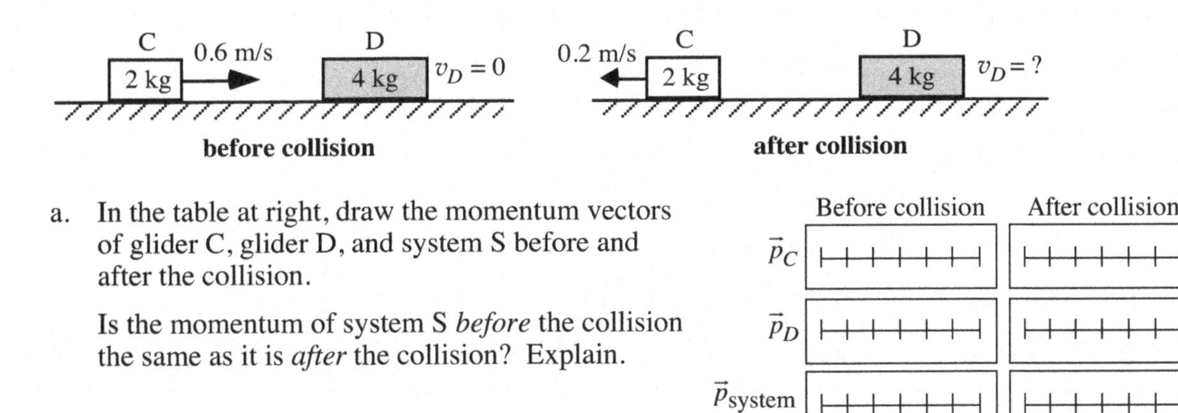

a. In the table at right, draw the momentum vectors of glider C, glider D, and system S before and after the collision.

Is the momentum of system S *before* the collision the same as it is *after* the collision? Explain.

b. Use your results from part a to determine the magnitude and direction of the velocity of glider D relative to the track after the collision. Explain.

Consider reference frame R, moving to the left with constant speed 0.2 m/s relative to the track.

c. Apply the Galilean transformation of velocities to draw the velocity vectors of gliders C and D *in reference frame R* before and after the collision.

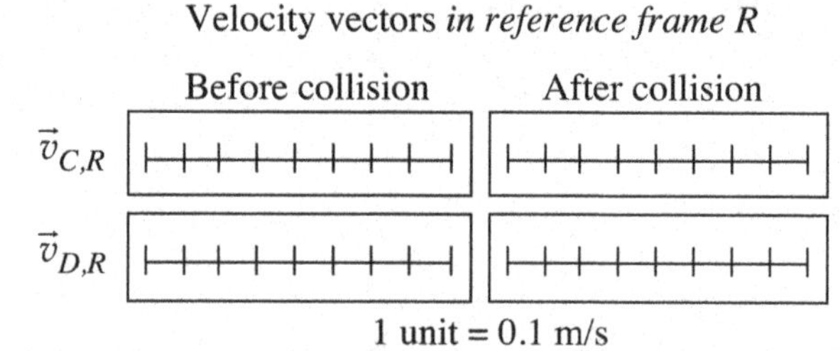

Use the velocity vectors above to draw the momentum vectors of glider C, glider D, and system S *in reference frame R* before and after the collision. Explain.

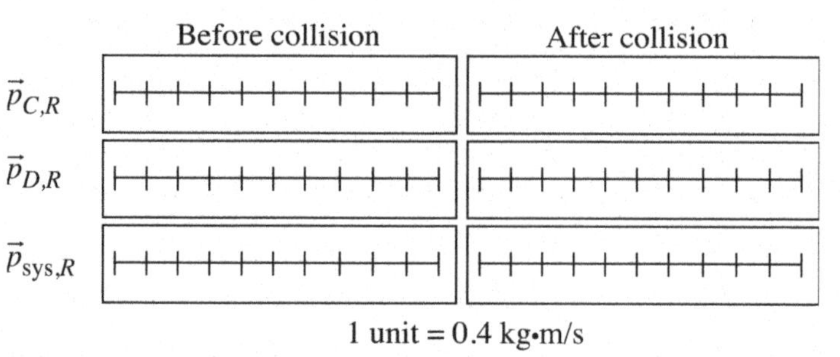

d. Use your results from part c above to answer the following:

i. In reference frame R, is the momentum of system S *before the collision* the same as it is *after the collision*?

ii. Recall the definition of *conserved* from part D of section II of the tutorial:

"When the momentum of an object or system of objects does not change with time, the momentum of the object or system is said to be *conserved*."

In reference frame R, is the momentum of system S conserved? Explain.

e. Use your results from parts a and c above to answer the following:

Before the collision, does the momentum of system S have the same value *in reference frame R* as it does *in the reference frame of the track?* Explain.

f. Consider the following dialogue.

Student 1: *"The momentum of system S is conserved. That means it must have the same value in reference frame R as it does in the frame of the track."*

Student 2: *"I disagree. Even when momentum is conserved, it could have a different value in one reference frame than it does in another. It must have the same value before and after the collision no matter which reference frame is used."*

With which student, if either, do you agree? Explain.

1. Two objects are arranged on a level, frictionless table as shown. Two experiments are conducted in which object A is launched toward the stationary block B. The initial speed of object A is the same in both experiments; the direction is not. The initial and final velocities of object A in each experiment are shown.

The mass of block B is four times that of object A $(m_B = 4m_A)$.

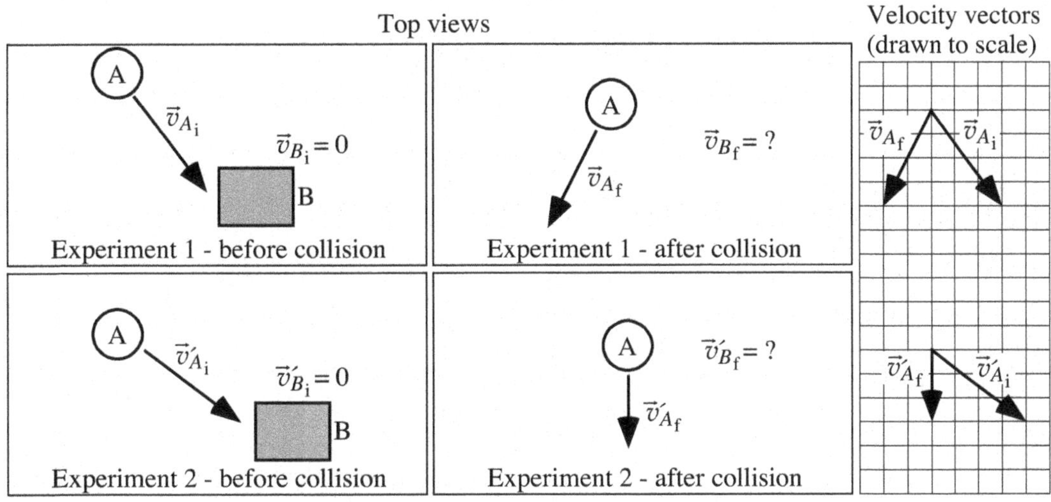

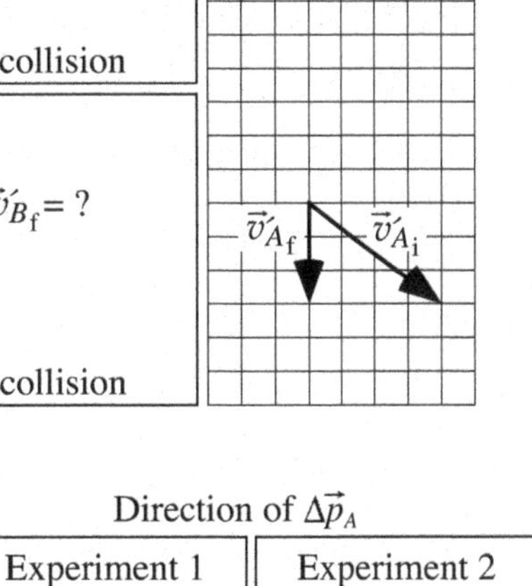

a. In the space provided, draw separate arrows representing the direction of the *change in momentum vector* of object A in the two experiments.

Direction of $\Delta\vec{p}_A$

Experiment 1	Experiment 2

Is the magnitude of the *change in momentum* of object A in experiment 1 *greater than, less than,* or *equal to* that in experiment 2? Explain.

b. In the space provided, draw separate arrows representing the direction of the *change in momentum vector* of block B in the two experiments.

Direction of $\Delta\vec{p}_B$

Experiment 1	Experiment 2

After the collisions, is the magnitude of the momentum of block B in experiment 1 *greater than, less than,* or *equal to* that in experiment 2? If the momentum of block B is zero in either case, state that explicitly. Explain.

2. Two objects collide on a level, frictionless table. The mass of object A is 5.0 kg; the mass of object B is 3.0 kg. The objects stick together after the collision. The initial velocity of object A and the final velocity of both objects are shown.

Before collision $\vec{v}_{A_i}$	After collision $(\vec{v}_{A_f} = \vec{v}_{B_f})$

(One side of a square represents 0.1 m/s)

a. In the space provided, draw separate arrows for object A and for object B representing the direction of the *change in momentum vector* of the object.

Direction of $\Delta\vec{p}$

Object A	Object B

Is the magnitude of the *change in momentum* of object A *greater than, less than, or equal to* that of object B? Explain your reasoning.

b. System C is the system of both objects A and B combined. How does the momentum of system C *before* the collision compare to the momentum of system C *after* the collision? Discuss both magnitude and direction.

$\vec{p}_{C_i}$

(Each side of a square represents 0.4 kg·m/s)

Construct and label a vector showing the momentum of system C at an instant before the collision. Show your work clearly.

c. Construct and label a vector showing the initial velocity of object B. Show your work clearly.

$\vec{v}_{B_i}$

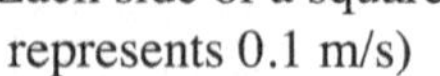

(Each side of a square represents 0.1 m/s)

3. a. Object A collides on a horizontal frictionless surface with an initially stationary target, object X. The initial and final velocities of object A are shown. The final velocity of object X is not given.

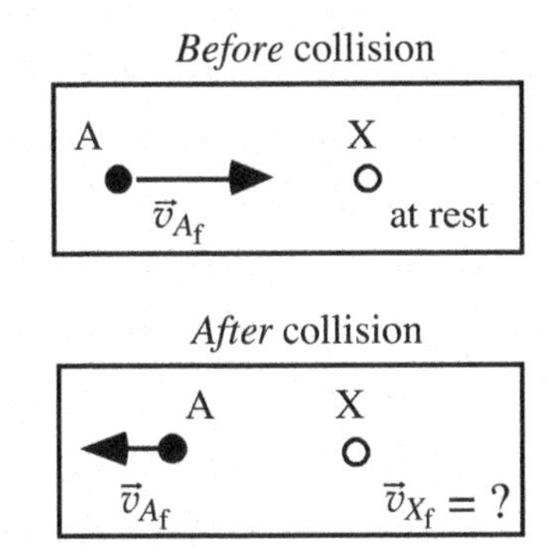

i. At an instant during the collision, is the net force on object A *zero* or *non-zero?*

ii. During the collision, is the momentum of object A conserved? Explain.

Is the momentum of the system consisting of objects A and X conserved? Explain.

b. On the same horizontal surface, object C collides with an initially stationary target, object Z. The initial speeds of objects C and A are the same, and $m_X = m_Z > m_A = m_C$.

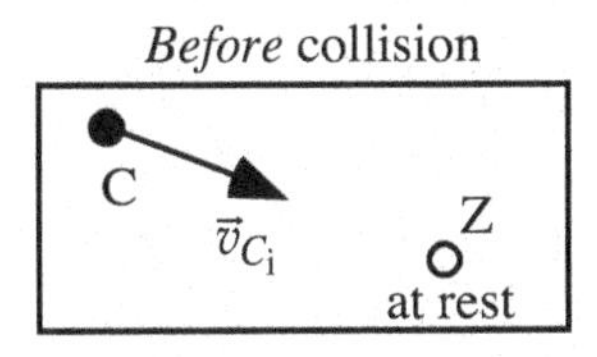

After the collisions, object C moves in the direction shown and has the same final speed as object A.

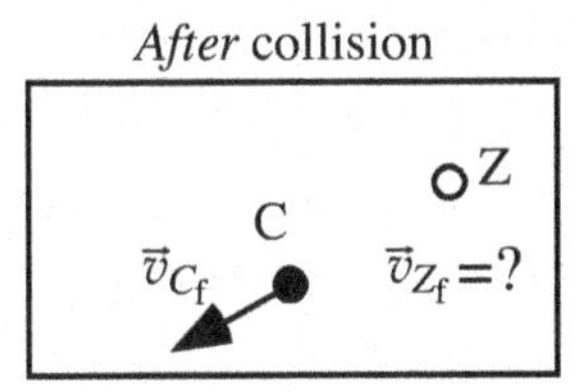

i. In the space below, copy the vectors $\vec{v}_{C_i}$ and $\vec{v}_{C_f}$ with their tails together. Use these vectors to draw the change in velocity vector $\Delta\vec{v}$ for glider C.

ii. Is the magnitude of the *change in velocity vector* of object A *greater than, less than,* or *equal to* the magnitude of the *change in velocity vector* of object C? Explain.

iii. Is the magnitude of the *change in momentum vector* of object A *greater than, less than,* or *equal to* the magnitude of the *change in momentum vector* of object C? Explain.

iv. Is the final speed of object X *greater than, less than,* or *equal to* the final speed of object Z? Explain.

c. Consider the following incorrect statement:

"Gliders A and C have the same change in momentum. They have the same mass, and because they have the same initial speed and same final speed, Δv is the same for each of them."

Discuss the error(s) in the reasoning.

1. A bicycle wheel is mounted on a fixed, *frictionless* axle. A light string is wound around the wheel's rim, and a weight is attached to the string at its free end. The diagrams below depict the situation at three different instants:

   At $t = t_o$, the weight is released from rest.

   At $t = t_1$, the weight is falling and the string is still partially wound around the wheel.

   At $t = t_2$, the weight and string have both reached the ground.

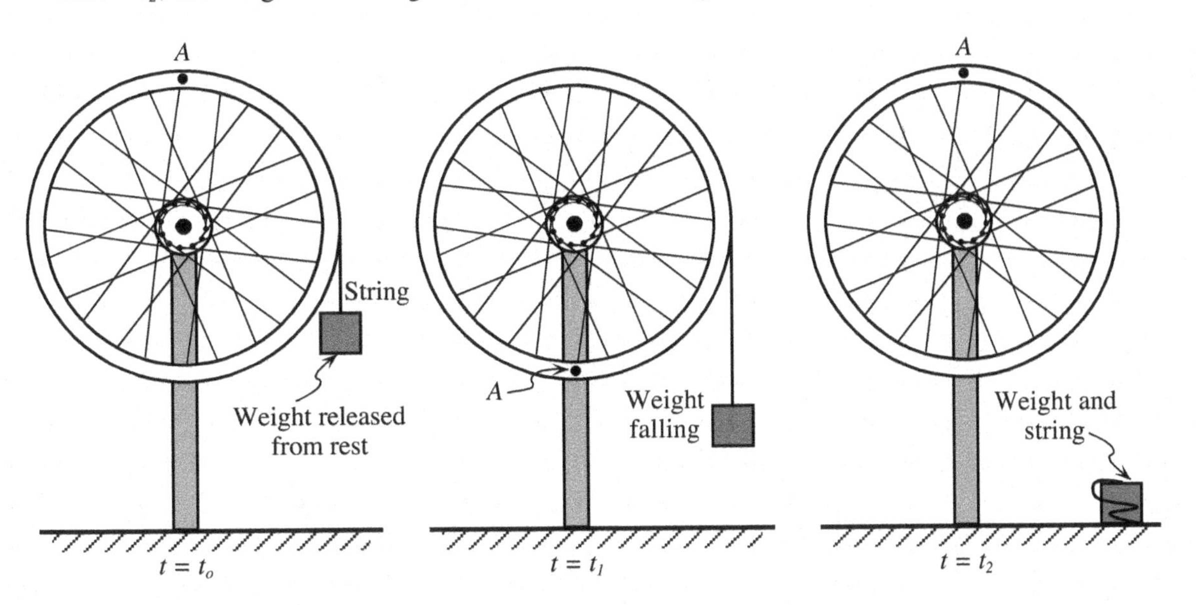

Note: The above diagrams do not represent stroboscopic photographs of the wheel.

a. What is the direction of the angular velocity $\vec{\omega}$ of the wheel at each time shown? If $\vec{\omega} = 0$ at any time, state that explicitly. Explain.

b. What is the direction of the wheel's angular acceleration $\vec{\alpha}$ at each time shown? If $\vec{\alpha} = 0$ at any time, state that explicitly. Explain.

c. Rank the magnitudes of the centripetal acceleration of point A at the three times shown $(a_{A_o}, a_{A_1}, a_{A_2})$. If any of these is zero, state that explicitly. Explain your reasoning.

2. A small piece of clay is stuck near the edge of a phonograph turntable. Let $\vec{\alpha}$ represent the angular acceleration of the clay and $\vec{a}_R$ represent the centripetal acceleration of the clay.

   a. For each situation described below, describe a possible motion of the turntable. (Some parts have more than one correct answer.) Explain your reasoning in each case.

      i. $\vec{\alpha} = 0$ and $\vec{a}_R = 0$

      ii. $\vec{\alpha} = 0$ and $\vec{a}_R \neq 0$

      iii. $\vec{\alpha} \neq 0$ and $\vec{a}_R \neq 0$

   b. The equation "$\vec{\alpha} = \vec{a}_R/r$" is *not* a correct vector equation.

      i. Explain how you can tell that the vector equation above is not correct.

      ii. Is "$\alpha = a_R/r$" a correct *scalar* equation? Explain why or why not.

1. A ruler is initially at rest and horizontal when a student briefly exerts a downward force on the right end. The magnitude of the force exerted by the student is less than the weight of the ruler.

   (Assume that the pivot is *frictionless*.)

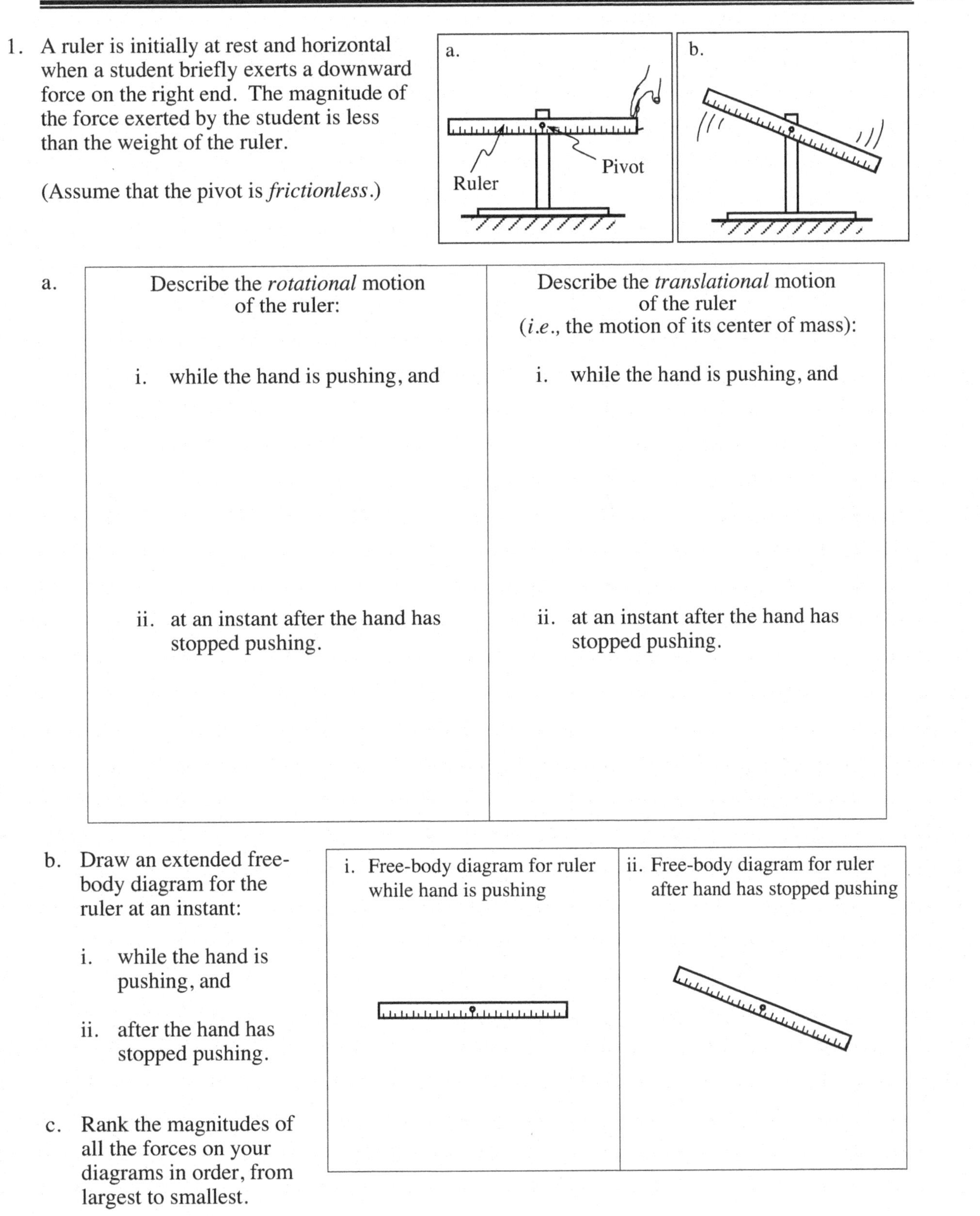

a.

Describe the *rotational* motion of the ruler:	Describe the *translational* motion of the ruler (*i.e.*, the motion of its center of mass):
i. while the hand is pushing, and	i. while the hand is pushing, and
ii. at an instant after the hand has stopped pushing.	ii. at an instant after the hand has stopped pushing.

b. Draw an extended free-body diagram for the ruler at an instant:

   i. while the hand is pushing, and

   ii. after the hand has stopped pushing.

i. Free-body diagram for ruler while hand is pushing	ii. Free-body diagram for ruler after hand has stopped pushing

c. Rank the magnitudes of all the forces on your diagrams in order, from largest to smallest.

d. Is the rotation of the ruler about the pivot *in a clockwise sense, in a counterclockwise sense,* or *zero* at each of the two instants shown? Explain.

Which of the forces on your free-body diagrams produce a non-zero torque? Explain.

e. In the boxes at right, indicate the direction of the net force on the ruler at each of the two instants in part b. If the net force is zero, state that explicitly. Explain.

While hand is pushing	After hand has stopped pushing

List all of the forces on each of your free-body diagrams that must be summed to obtain the net force. Explain.

f. Check that your descriptions of the motion of the ruler in part a are consistent with your answers in parts d and e.

2. Three identical rectangular blocks are at rest on a horizontal, frictionless ice rink. Forces of equal magnitude and direction are exerted on each of the three blocks. Each force is exerted at a different point on the block, as shown in the top view diagram below. (Each "×" in the diagram indicates the location of the block's center of mass.)

a. Parts i–iv refer to the instant shown in the diagram below.

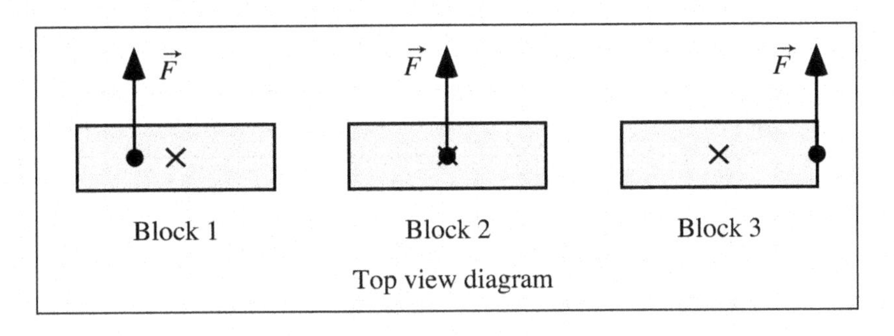

Top view diagram

i. For each block, state whether the *angular acceleration* is *in a clockwise sense, in a counterclockwise sense,* or *zero*. Explain.

ii. Rank the magnitudes of the angular accelerations of the blocks (α_1, α_2, α_3). Explain.

iii. The diagrams on the previous page have been reproduced below. Sketch a vector on each block to indicate the direction of the acceleration of the center of mass ($\vec{a}_{cm}$) of that block. If for any block $\vec{a}_{cm} = 0$, state that explicitly. Explain.

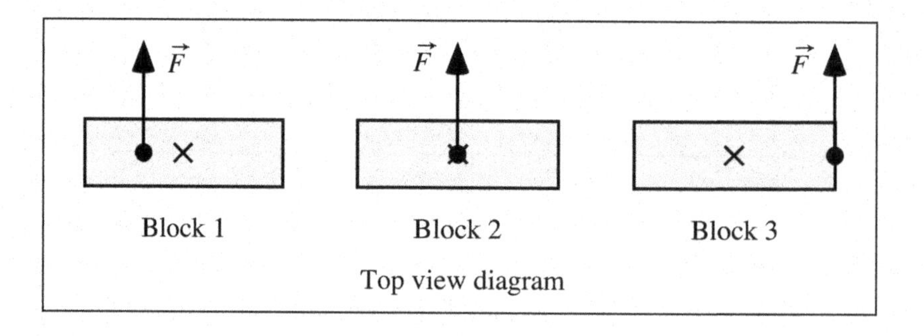

iv. Rank the magnitudes of the accelerations of the centers of mass of the blocks ($a_{cm,1}$, $a_{cm,2}$, $a_{cm,3}$). Explain.

b. Suppose that each force shown above is exerted for the same time interval Δt.

Compare the *magnitudes* of the final *angular velocities* of the blocks (ω_{1f}, ω_{2f}, ω_{3f}) at the end of this time interval. If for any block $\omega_f = 0$, state that explicitly. Explain.

1. Recall that in part C of section II of the tutorial *Equilibrium of rigid bodies* a T-shaped board was hung from the hole through the center of mass and held at an angle as shown. You observed that the board remained at rest when released.

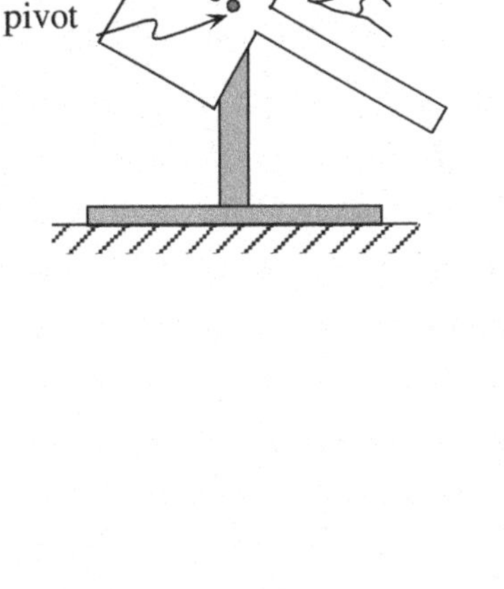

   a. What does this observation imply about the net torque about the pivot? Explain.

   b. Draw an extended free-body diagram for the board in the space at right.

   Extended free-body diagram for board

   Is the point at which you placed the gravitational force on your diagram consistent with your answer for the net torque about the pivot? Explain.

2. Suppose the board and piece of clay were in equilibrium as in section II of the tutorial *Equilibrium of rigid bodies*. Imagine that the piece of clay is moved to a location closer to the pivot by moving it directly upward to point *B*. Neglect friction.

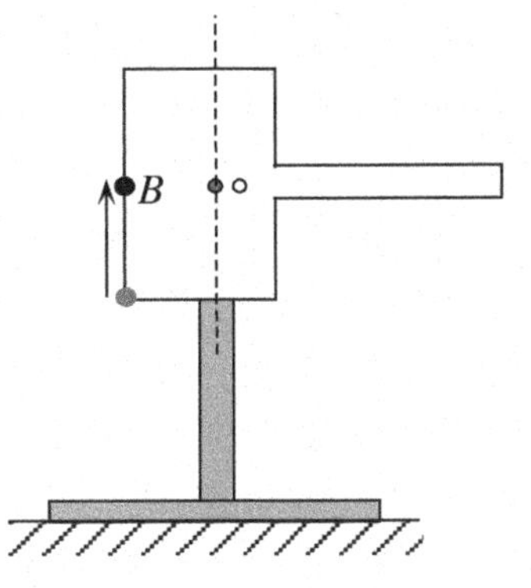

   a. Would the board remain at rest in this case? Explain.

   b. On the figure above, mark the approximate location of the center of mass of the system composed of clay (at point *B*) and board with an "x." Explain.

   c. Is the amount of mass of the system (clay and board) to the left of the dashed line *greater than, less than,* or *equal to* the amount of mass of the system to the right of the dashed line? Explain.

3. a. A T-shaped sheet of uniform thickness has a uniform mass density.

Is the center of mass of the sheet located *above point R, below point R,* or *at point R?* Explain your reasoning. (*Hint:* Do not assume that this board is identical to the one used in the tutorial *Equilibrium of rigid bodies*.)

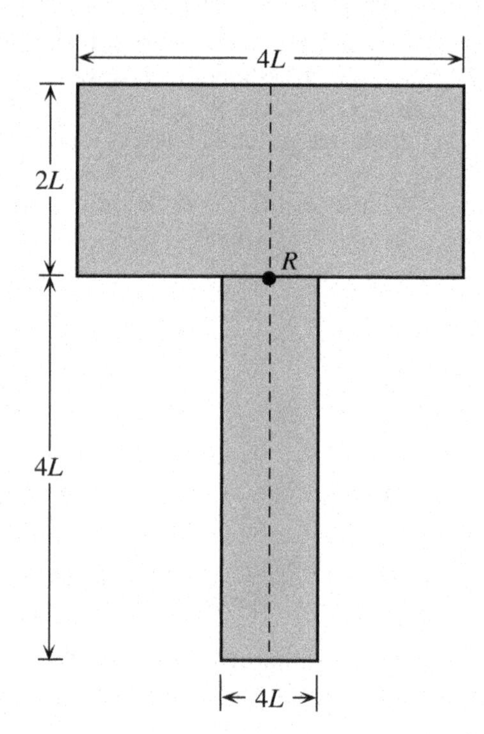

b. A different T-shaped sheet of uniform thickness is composed of two different materials designated by the shaded and unshaded regions in the diagram at right. The mass density of *each piece* is uniform throughout its volume. The mass density of the shaded piece is twice as large as the mass density of the unshaded piece.

Is the center of mass of the sheet located *above point R, below point R,* or *at point R?* Explain your reasoning.

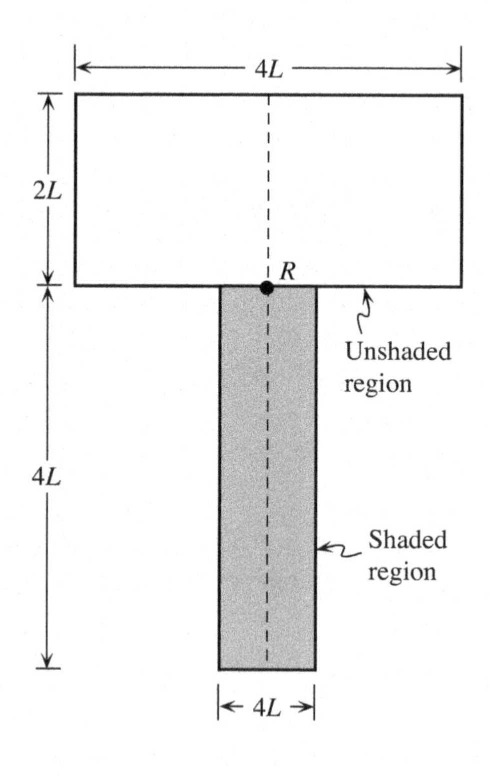

Electricity and magnetism

1. Two identical metal balls are suspended by insulating threads. Both balls have the same net charge. In this problem, do not assume the balls are point charges.

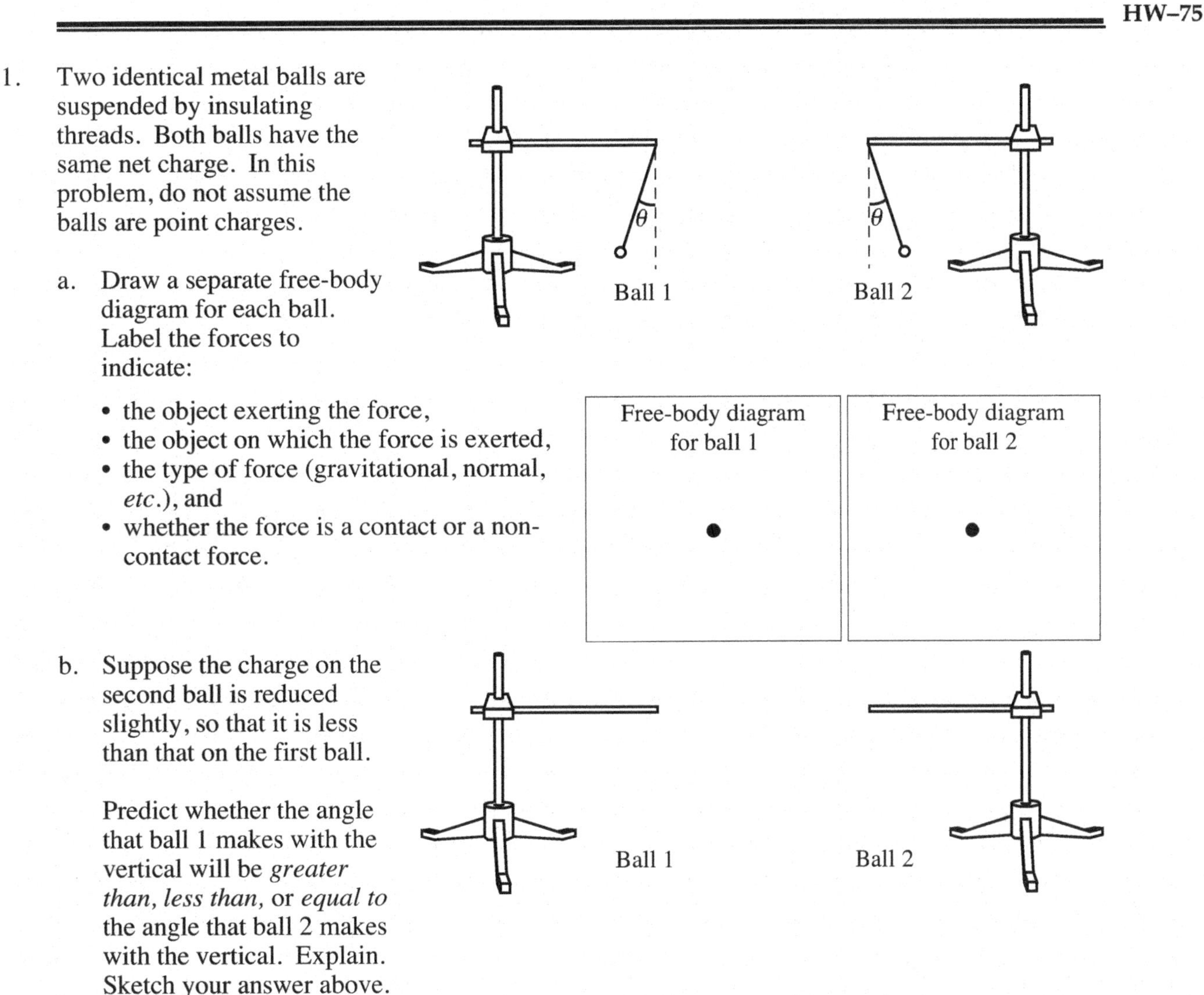

   a. Draw a separate free-body diagram for each ball. Label the forces to indicate:

   - the object exerting the force,
   - the object on which the force is exerted,
   - the type of force (gravitational, normal, *etc*.), and
   - whether the force is a contact or a non-contact force.

   b. Suppose the charge on the second ball is reduced slightly, so that it is less than that on the first ball.

   Predict whether the angle that ball 1 makes with the vertical will be *greater than, less than,* or *equal to* the angle that ball 2 makes with the vertical. Explain. Sketch your answer above.

   How does the free-body diagram for each ball in this case compare to the corresponding free-body diagram that you drew in part a? If the magnitudes or directions of any of the forces change, describe how they change.

   c. Predict what will happen if the net charge on ball 2 is reduced to zero. Make a sketch to illustrate your answer.

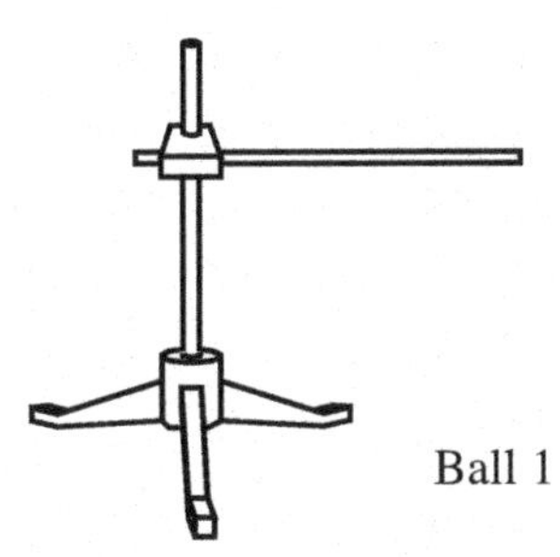

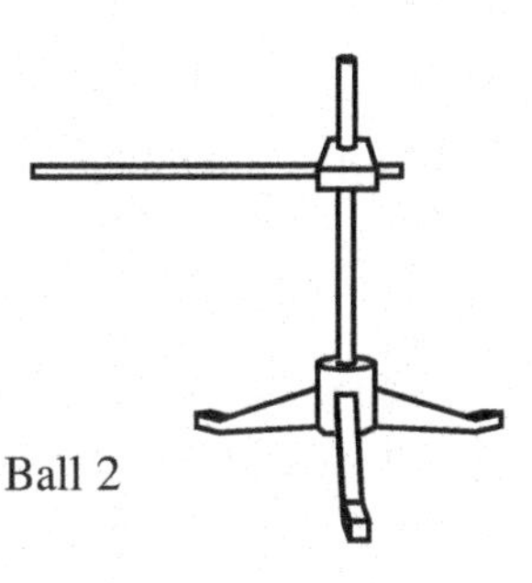

2. *Coulomb's law* allows us to find the force between two *point* charges.

   Three point charges are held fixed in place as shown.

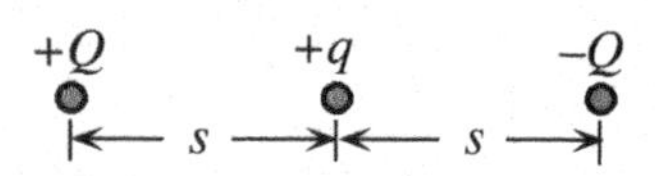

   Consider the following comment about this situation:

   "There will be zero net electric force on the charge in the middle due to the other charges. Using Coulomb's law, the force due to the +Q charge is positive, and the force due to the -Q charge is negative. The forces cancel."

   a. Do you agree with this statement? Explain.

   b. How does Coulomb's law apply to situations in which there are more than two point charges?

3. Each of the following parts involves a comparison of the net electric force exerted on a positive charge $+q$ in two different cases.

   a. In cases A and B shown at right there are two positive point charges $+Q$ each a distance s away from a third positive point charge $+q$.

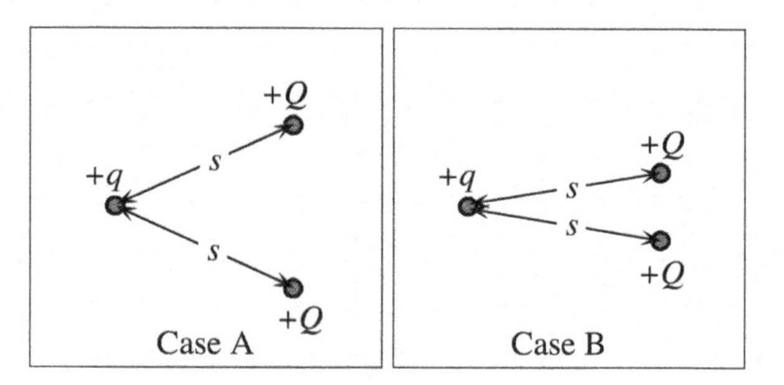

   Is the net electric force on the $+q$ charge in case A *greater than, less than,* or *equal to* the net electric force on the $+q$ charge in case B? Explain.

   b. In case C, two positive point charges $+2Q$ are each a distance s away from a third positive point charge $+q$. In case D, four positive point charges $+Q$ are each a distance s away from a fifth positive point charge $+q$. (The angle α shown is the same in both cases.)

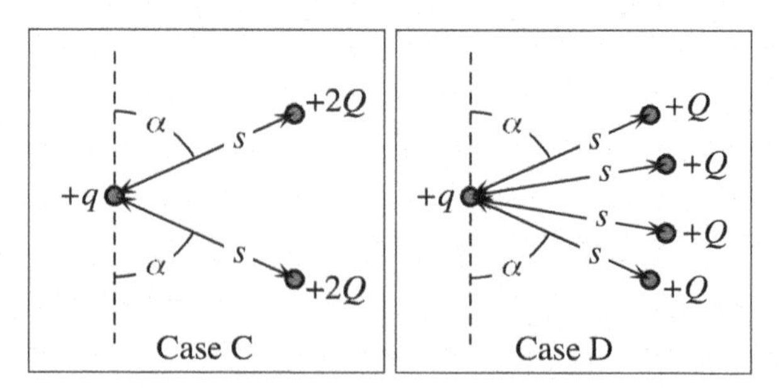

   Is the net electric force on the $+q$ charge in case C *greater than, less than,* or *equal to* the net electric force on the $+q$ charge in case D? Explain.

c. In case E a positive point charge with $+Q$ is a distance s away from a third positive point charge $+q$. In case F ten positive point charges, each with charge $+Q/10$, lie along an arc of radius s centered on a positive point charge $+q$.

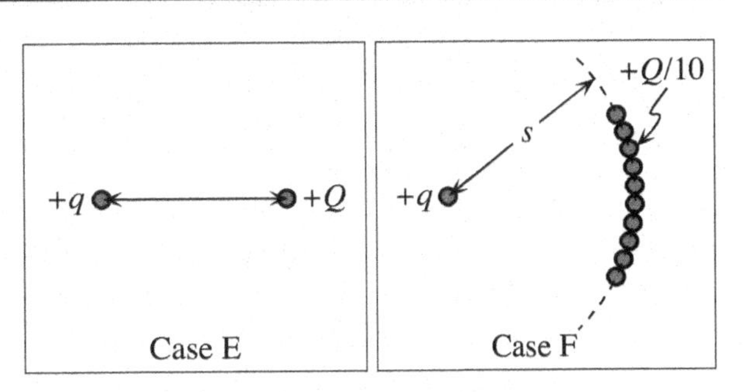

Is the net electric force on the $+q$ charge in case E *greater than, less than,* or *equal to* the net electric force on the $+q$ charge in case F? Explain.

4. A thin semicircular rod has a total charge $+Q$ uniformly distributed along it. A negative point charge $-Q$ is placed as shown. A test charge $+q$ is placed at point C. (Point C is equidistant from $-Q$ and from all points on the rod.)

Let F_P and F_R represent the force on the test charge due to the point charge and the rod respectively.

a. Is the magnitude of F_P *greater than, less than,* or *equal to* the magnitude of F_R? Explain how you can tell.

b. Is the magnitude of the *net* force on $+q$ *greater than, less than,* or *equal to* the magnitude of F_P? Explain.

c. A second negative point charge $-Q$ is placed as shown.

Is the magnitude of the *net* electric force on $+q$ *greater than, less than,* or *equal to* the magnitude of the net electric force on $+q$ in part b? Explain.

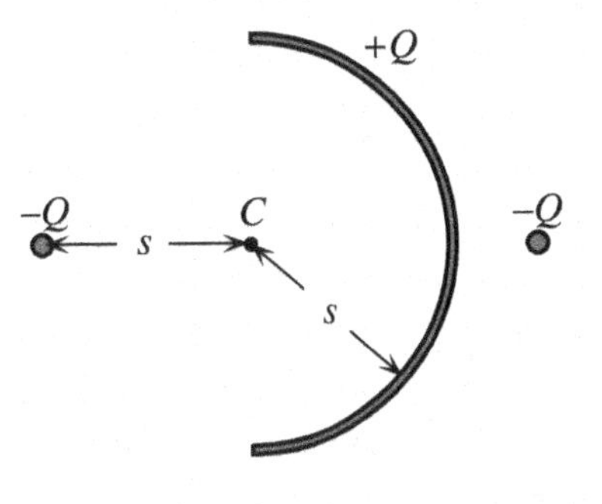

5. A thin semicircular rod like the one in problem 4 is broken into two halves. The top half has a total charge $+Q$ uniformly distributed along it, and the bottom half has a total charge $-Q$ uniformly distributed along it.

   On the diagram, indicate the direction of the net electric force on a positive test charge placed in turn at points *A*, *B*, and *C*. Explain how you determined your answers.

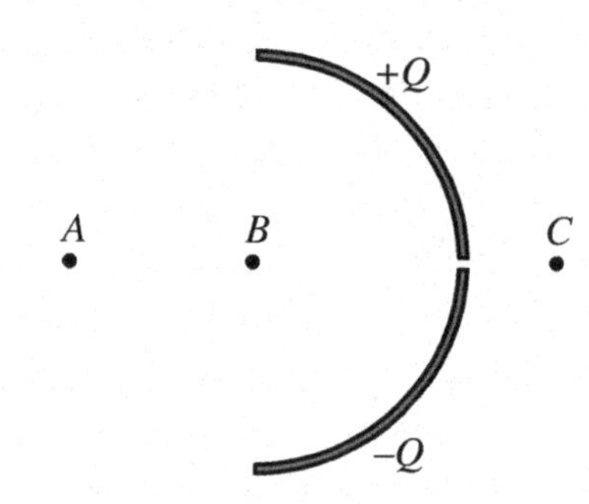

6. A positive point charge $+q$ is placed near an uncharged metal rod.

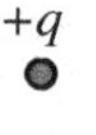

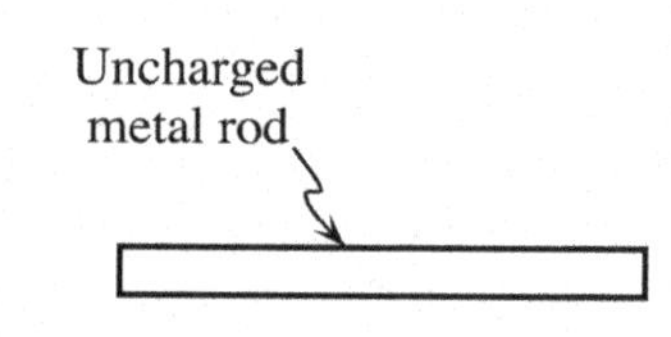

   a. Sketch the charge distribution on the rod.

   b. Is there a non-zero net electric force on the rod? Explain.

   c. Is there a non-zero net electric force on the point charge? Explain.

7. State whether the magnitude of the net electric force on the charge labeled $+Q_o$ in case A is *greater than, less than,* or *equal to* the magnitude of the net electric force on the charge labeled $+Q_o$ in case B. Explain how you determined your answer.

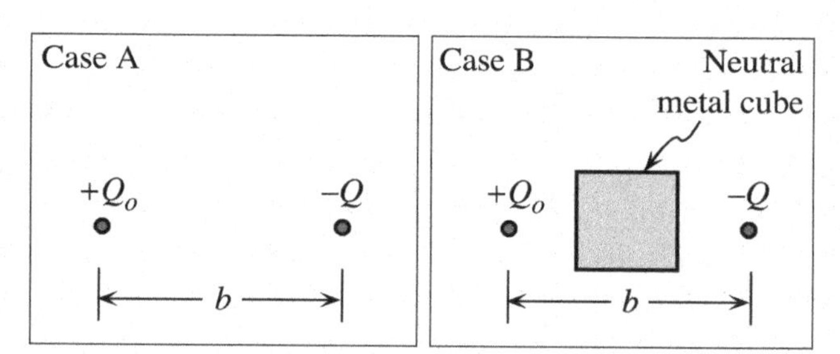

ELECTRIC FIELD AND FLUX

Name ____________________

1. A piece of paper is folded into three equal parts as shown.

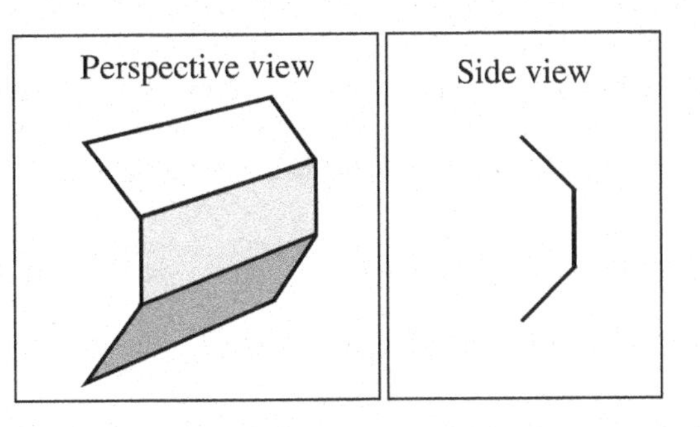

   a. How many area vectors are needed to describe the surface of the paper? Sketch the area vectors on the side view diagram and label them $\vec{A}_1$, $\vec{A}_2$, *etc.*

   b. Consider an imaginary surface in a uniform electric field $\vec{E}$ as shown. The surface has the same size and shape as the paper above. Is the flux through the top third of the surface *greater than, less than,* or *equal to* the flux through the middle third? Explain.

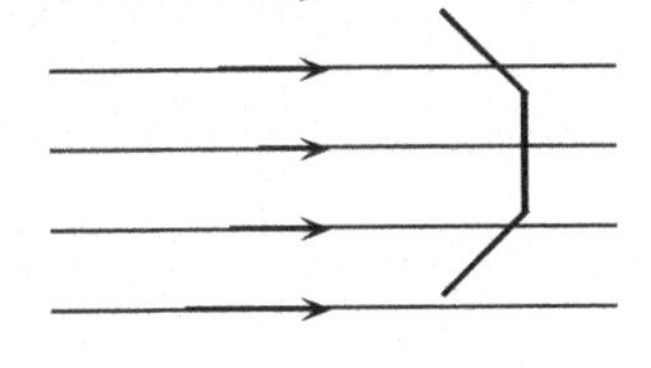

   c. Write an expression for the net electric flux Φ_{net} through the entire surface in terms of the area vectors and the electric field $\vec{E}$. (*Hint:* Use the vector definition for electric flux found in tutorial to first write expressions for the flux through each of the flat surfaces.)

2. A positive charge is located at the center of a cube.

    a. Are the intersections of the field lines with a side of the box uniformly distributed across that side? Explain.

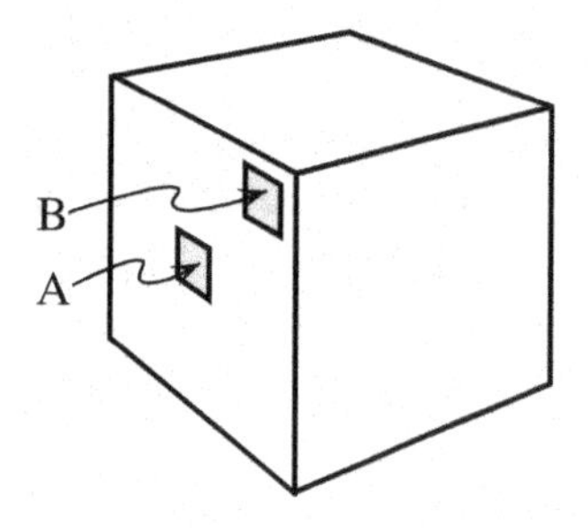

    b. We can consider the left side of the box as composed of many small surface elements of equal area.

        i. Is the number of field lines through surface element A *greater than, less than,* or *equal to* the number of field lines through surface element B? Explain.

        ii. Is the flux through surface element A *greater than, less than,* or *equal to* the flux through surface element B? Explain.

    c. Consider the surface element A itself as composed of many even smaller pieces. Would the number of field lines through each of those new small surface elements vary much from one to another? Explain.

    Describe how the field lines for the positive point charge appear to be distributed when the region over which you look becomes sufficiently small.

    d. Consider the left side of the box as consisting of N small pieces. Let $d\vec{A}_i$ represent the area of the ith small surface element on the left side of the box, and let $\vec{E}_i$ represent the electric field on that surface element.

    Write an expression for the net electric flux Φ_{net} through the left side of the box in terms of $d\vec{A}_i$ and $\vec{E}_i$.

3. The loop shown at right has an area A.

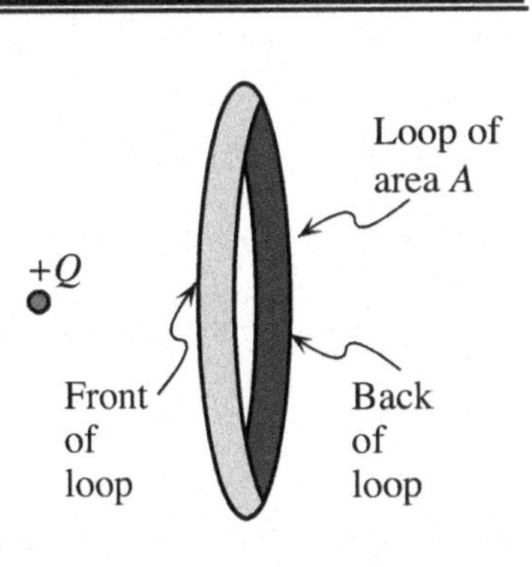

a. The loop is held to the right of a positive point charge as shown.

i. Draw and label an area vector for the surface bounded by the loop.

ii. Sketch electric field lines to represent the electric field due to the charge.

iii. Is the electric flux through the loop due to the charge *positive, negative,* or *zero?* Explain your reasoning.

b. A positive charge with twice the value of the initial charge is now placed to the right of the loop as shown. Both charges are the same distance from the loop and are placed along the axis of the loop.

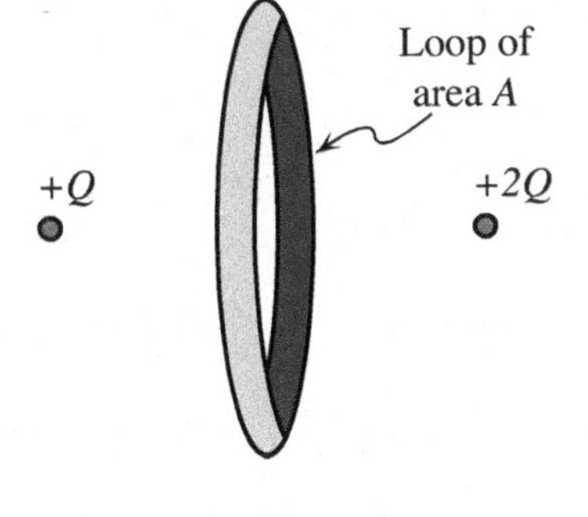

Is the net electric flux through the loop due to the charges *positive, negative,* or *zero?* Explain your reasoning. (*Note:* Use the same area vector you used in part a.)

c. Suppose that the new charge located to the right of the loop had been negative instead of positive. How would your answer to part b change, if at all? Explain.

1. The *closed* Gaussian surface shown at right consists of a hemispherical surface and a flat plane. A point charge $+q$ is outside the surface, and no charge is enclosed by the surface.

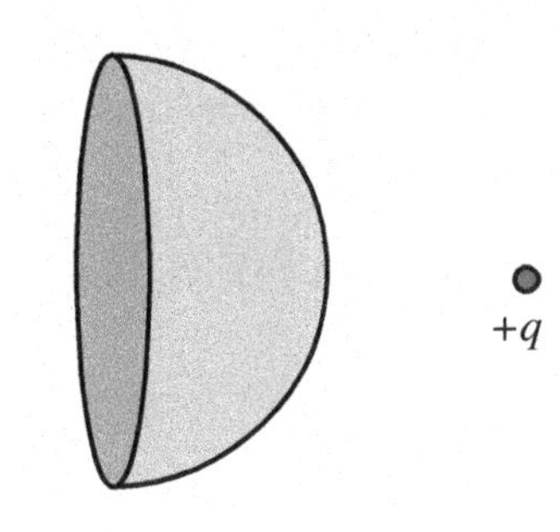

   a. What is the flux through the *entire* closed surface? Explain.

   Let Φ_L represent the flux through the flat left-hand portion of the surface. Write an expression in terms of Φ_L for the flux through the curved portion of the surface, Φ_C.

   b. Suppose that the curved portion of the Gaussian surface in part a is replaced by the larger curved surface as shown. The flat left-hand portion of the surface is unchanged.

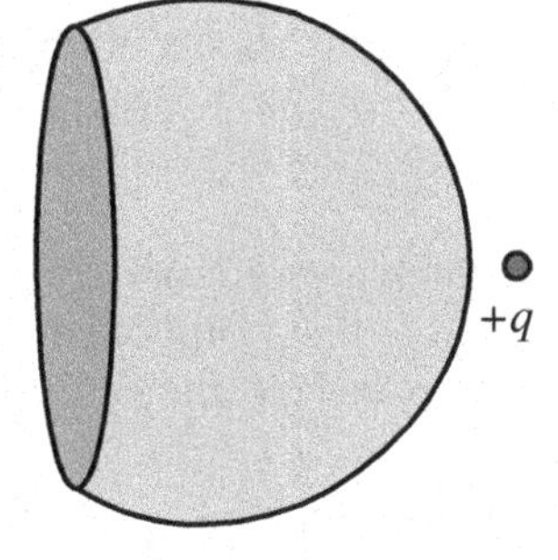

      i. Does the value of Φ_L change? Explain.

      ii. How does the flux through the new curved portion of the surface compare to the flux through the original curved portion of the surface? Explain.

   c. Suppose that the curved portion of the Gaussian surface is replaced by the larger curved surface that encloses the charge as shown. The flat left-hand portion of the surface is still unchanged.

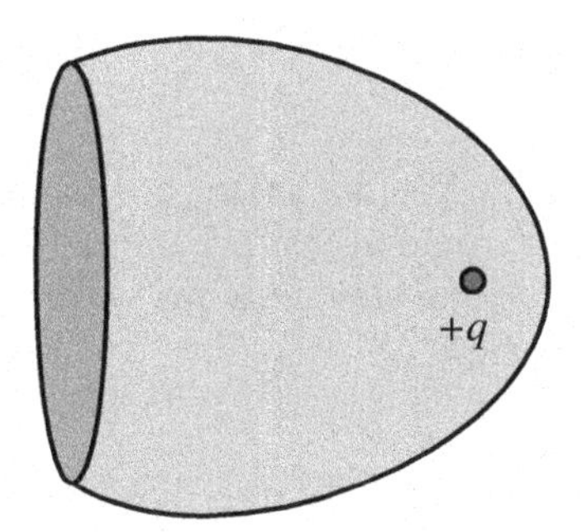

      i. Does the value of Φ_L change? Explain.

      ii. How does the flux through the new curved portion of the surface compare to the flux through the original curved portion of the surface? Explain.

      iii. Use Gauss' law to write an expression in terms of Φ_L and q for the flux through the curved portion of the surface.

d. A second point charge $+q$ is placed to the right of the Gaussian surface as shown.

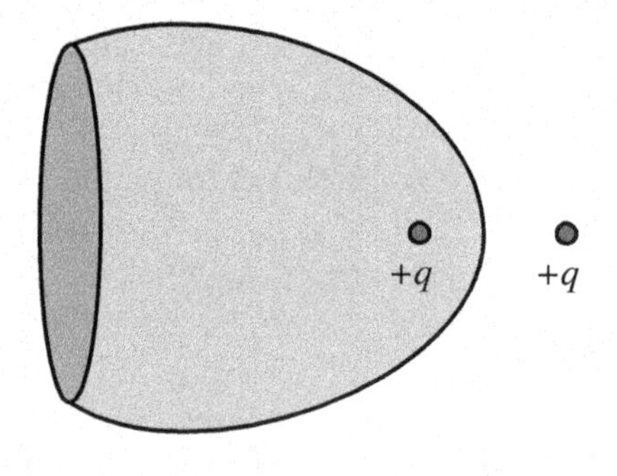

i. Is the value of Φ_L *greater than, less than,* or *equal to* the value of Φ_L in part c? Explain.

ii. Is the value of the flux through the entire Gaussian surface *greater than, less than,* or *equal to* the value of the flux through the entire Gaussian surface in part c? Explain.

2. Consider three sheets of charge with the charge densities shown. The sheets are very large and extend beyond the top and bottom of the side view diagram at right.

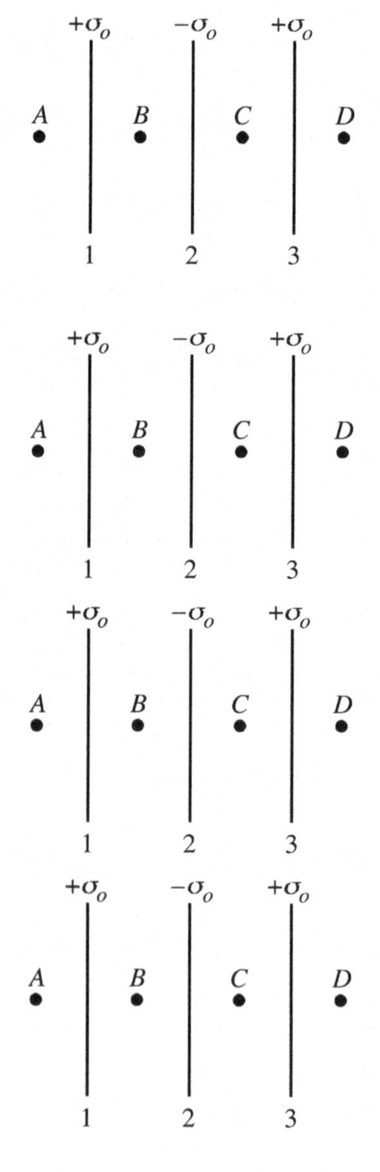

a. Sketch a vector at each of points *A–D* to represent the electric field at that point due to sheet 1. Draw your vectors to scale and state how they compare in magnitude.

b. Sketch a vector at each of points *A–D* to represent the electric field at that point due to sheet 2. Draw your vectors to scale and state how they compare in magnitude.

c. Sketch a vector at each of points *A–D* to represent the electric field at that point due to sheet 3. Draw your vectors to scale and state how they compare in magnitude.

d. Sketch the net electric field at each of points *A–D*.

e. Calculate the magnitude of the electric field at each point *A–D*. Use superposition to answer this question.

1. Two charged rods, each with net charge $-Q_o$, are held in place as shown in the top view diagram below.

   a. A small test charge $-q_o$ travels from point X to point Y along the circular arc shown.

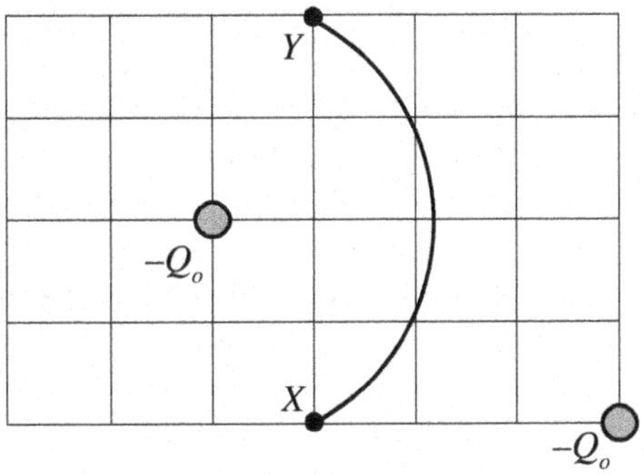

      i. Draw an arrow on the diagram at each point (X and Y) to show the direction of the electric force on the test charge at that point. Explain why you drew the arrows as you did.

      Y $-Q_o$ X $-Q_o$

      ii. Is the work done on the charge by the electric field *positive, negative,* or *zero?* Explain.

      iii. Is the electric potential difference ΔV_{XY} *positive, negative,* or *zero?* Explain.

   b. The test charge is launched from point X with an initial speed v_o and is observed to pass through point Y. Is the speed of the test charge at point Y *greater than, less than,* or *equal to* v_o? Explain your reasoning.

2. A positive charge of magnitude q_o is shown in the diagram below.

   a. Points B and C are a distance r_o away from the charge and point A is a distance $2r_o$ from it.

   $+q_o$

   × A × B ●

   × C

      i. Indicate the direction of the electric field at points A, B, and C on the diagram.

      ii. Let W_{AB} represent the absolute value of the work done by an external agent in moving a small test charge from point A to point B.

         • Would the absolute value of the work done by an external agent in moving the same test charge from point B to point C be *greater than, less than,* or *equal to* W_{AB}? Explain.

         • Would the absolute value of the work done by an external agent in moving the same test charge from point A to point C be *greater than, less than,* or *equal to* W_{AB}? Explain.

   b. A large metal sphere with zero net charge is now placed to the left of point A as shown.

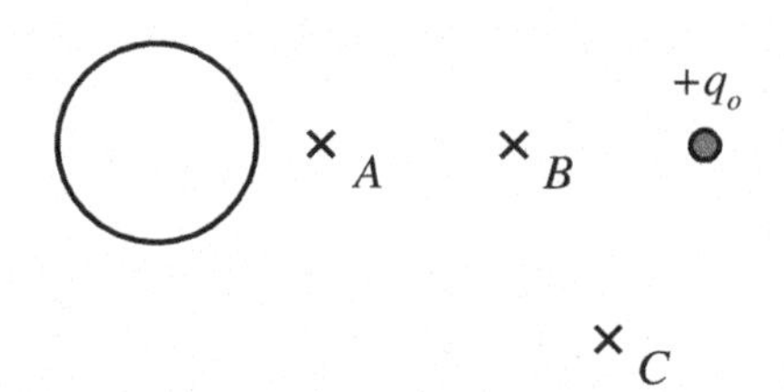

      i. Sketch the charge distribution on the metal sphere in the diagram at right.

      ii. Has the *magnitude* of the electric field at the following points *increased, decreased,* or *remained the same?* Explain.

         • point B

         • point C

iii. Has the *direction* of the electric field at the following points changed? Explain.

- point B

- point C

iv. Has the absolute value of the electric potential difference ΔV_{AB} from point A to point B *increased, decreased,* or *remained the same?* Explain your reasoning.

3. Two very large sheets of charge are separated by a distance d. One sheet has a surface charge density $+\sigma_o$ and the other a surface charge density $-\sigma_o$. A small region near the center of the sheets is shown.

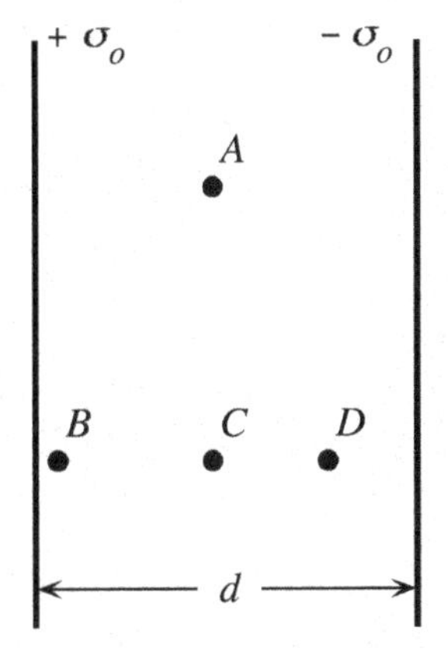

a. Draw arrows on the diagram to indicate the direction of the electric field at points A, B, C, and D.

i. Compare the magnitudes of the electric fields at points A, B, C, and D. Explain.

ii. How would the electric force exerted on a charged particle at point A compare to the electric force exerted on the same particle at point B? point C? point D?

b. A positively charged test particle moves from point *A* to point *C*.

i. Is the work done on the particle by the electric field *positive, negative,* or *zero?*

ii. Find ΔV_{AC}. Explain how you found your answer.

c. A positively charged test particle moves from point *A* to point *D*.

i. Is the work done on the particle by the electric field *positive, negative,* or *zero?*

ii. Is ΔV_{AD} *positive, negative,* or *zero?* Explain how you can tell.

d. Find the magnitude and direction of the electric field at points *A, B, C,* and *D*. (*Hint:* Use superposition and your results for the electric field of a large sheet from the *Gauss' law* tutorial.)

e. A particle of mass m_o and charge $-q_o$ is released from rest at a point just to the left of the right sheet. Find the speed of the particle as it reaches the left sheet.

1. Two large flat plates are separated by a distance d. The plates are connected to a battery.

   a. The surface area of the face of each plate is A_1.

      Write an expression for the capacitance in terms of A_1 and d.

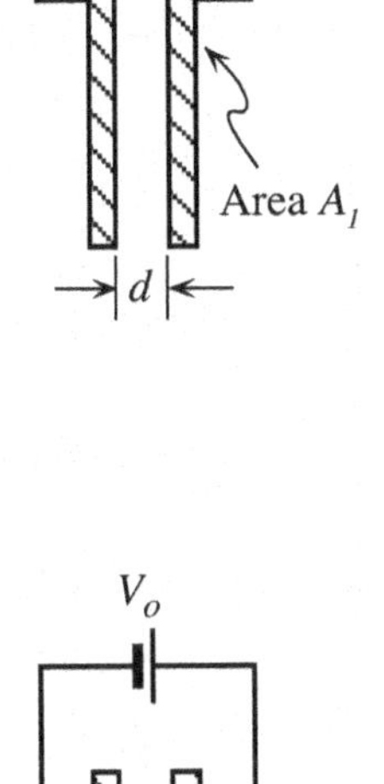

   b. A new capacitor is formed by attaching two uncharged metal plates, each with area A_2, to the capacitor as shown. The battery remains connected.

      i. When the new plates are attached, does the electric potential difference between the plates *increase, decrease,* or *remain the same?* Explain.

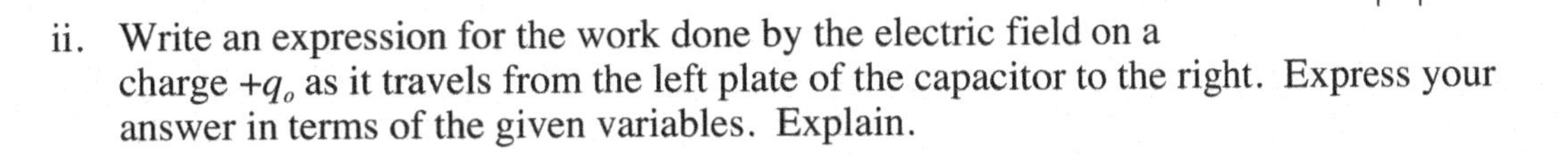

      ii. Write an expression for the work done by the electric field on a charge $+q_o$ as it travels from the left plate of the capacitor to the right. Express your answer in terms of the given variables. Explain.

      iii. Write an expression for the magnitude and direction of the electric field between the plates. Is the magnitude of the electric field *greater than, less than,* or *equal to* the magnitude of the electric field between the plates before the new plates were attached?

      iv. Write an expression for the charge density on the plates of the capacitor. Is the charge density *greater than, less than,* or *equal to* the charge density on the plates before the new plates were attached? Explain.

v. Write an expression for the total charge on one of the plates of the capacitor. Is this total charge *greater than, less than,* or *equal to* the total charge on one of the original plates? Explain.

vi. Use the definition of capacitance to find the capacitance of the enlarged pair of plates. Has the capacitance *increased, decreased,* or *remained the same?*

2. Two plates with area A_1 are held a distance d apart and have a net charge $+Q_1$ and $-Q_1$, respectively. Assume that all the charge is uniformly distributed on the inner surfaces of the plates.

Two initially uncharged plates of surface area A_2 are then attached to the original plates as shown.

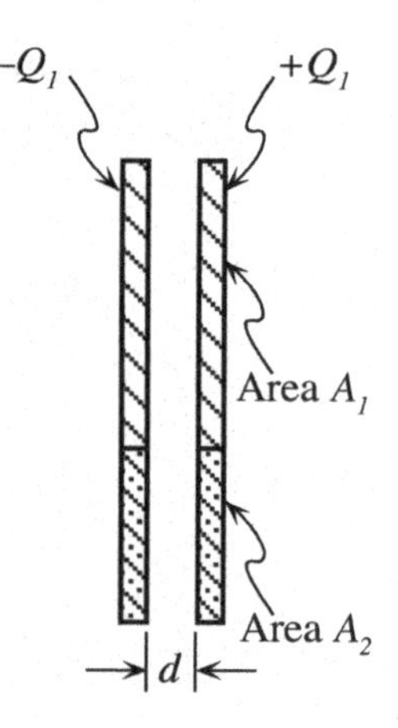

a. Find the charge density on the plates. Explain.

b. Find the electric potential difference between the plates. Explain.

c. Show that the capacitance of the enlarged plates in this case is the same as the capacitance you found in problem 1 of this homework.

1. In tutorial, you compared the relative brightness of the bulbs in the three circuits shown. In the diagrams, boxes have been drawn around the networks of bulbs in each circuit.

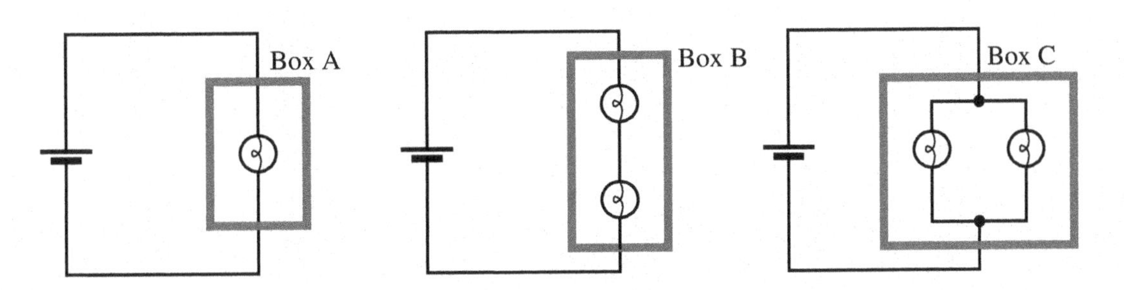

On the basis of your observations and the rule you developed in tutorial relating current through the battery to total resistance, rank the networks (boxes) A–C according to their equivalent resistance. Explain your reasoning on the basis of the model. (Do not use math.)

2. Use the model for electric current to rank the networks shown below in order according to resistance. Explain your reasoning.

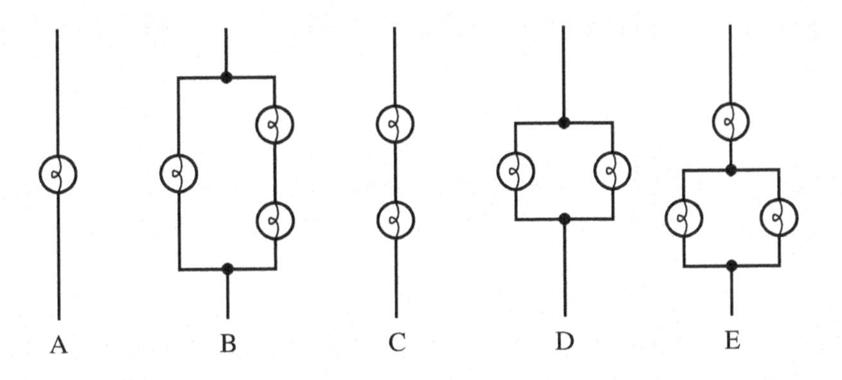

3. The circuit shown has four identical light bulbs and an ideal battery.

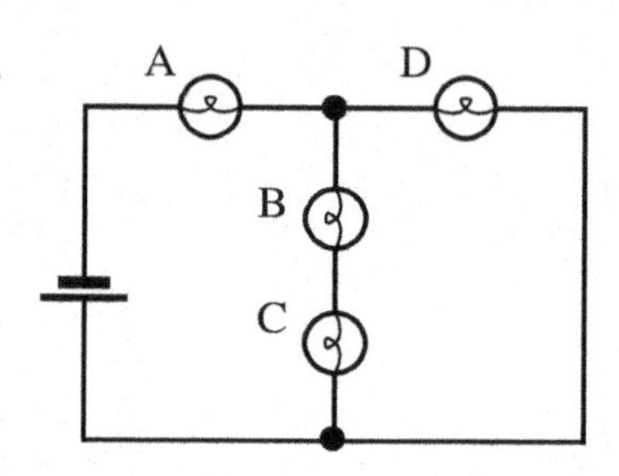

   a. Rank the brightness of the bulbs. Explain your reasoning.

   b. A wire is now added to the circuit as shown.

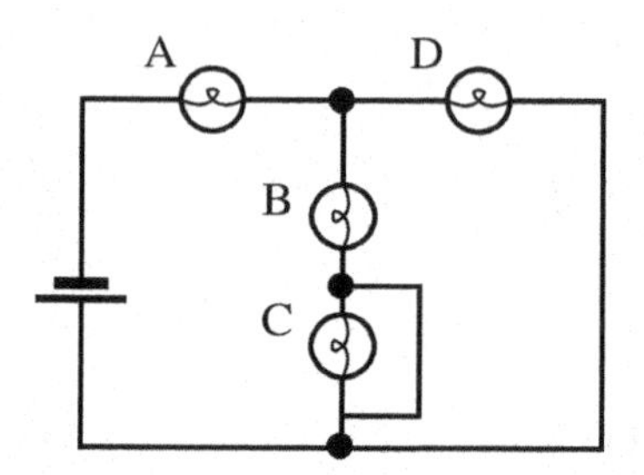

      i. Does the brightness of bulb C *increase, decrease,* or *remain the same?* Explain your reasoning.

      ii. Does the brightness of bulb A *increase, decrease,* or *remain the same?* Explain.

      iii. Does the current through the battery *increase, decrease,* or *remain the same?* Explain.

Name ______________________

4. Consider the five networks shown at right.

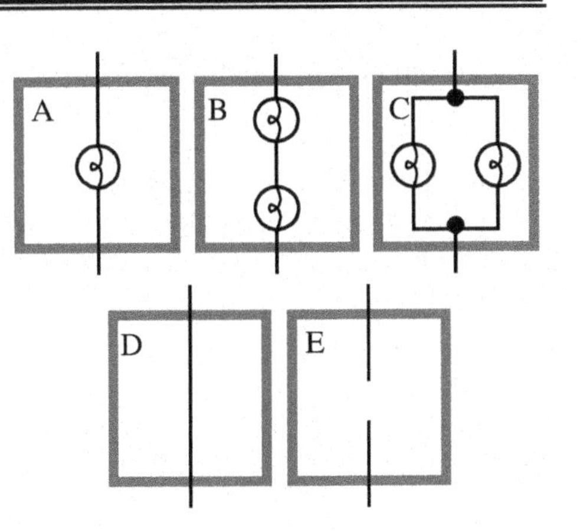

a. Rank the networks according to their equivalent resistance. (*Hint:* Imagine placing each network in series with an indicator bulb and a battery.)

b. How does adding a single bulb to a circuit in *series* with another bulb or network affect the resistance of the circuit?

c. How does adding a single bulb to a circuit in *parallel* with another bulb or network affect the resistance of the circuit?

d. The networks A–E above are connected, in turn, to identical batteries as shown. Use the model we have developed to:

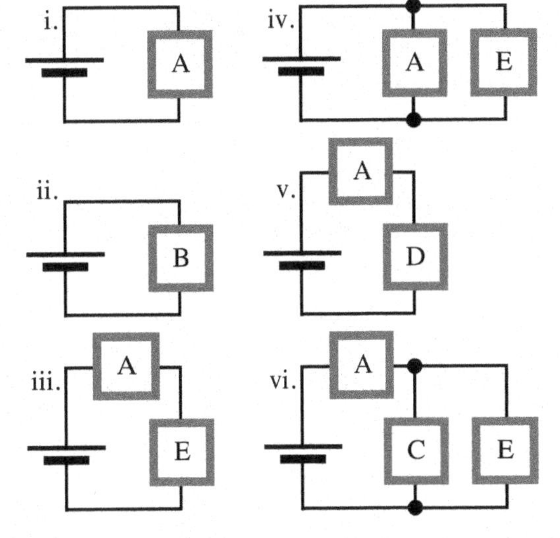

- rank the circuits according to equivalent resistance. Explain.

- rank the circuits according to the current through the battery. Explain.

5. The circuit below shows four identical bulbs connected to an ideal battery.

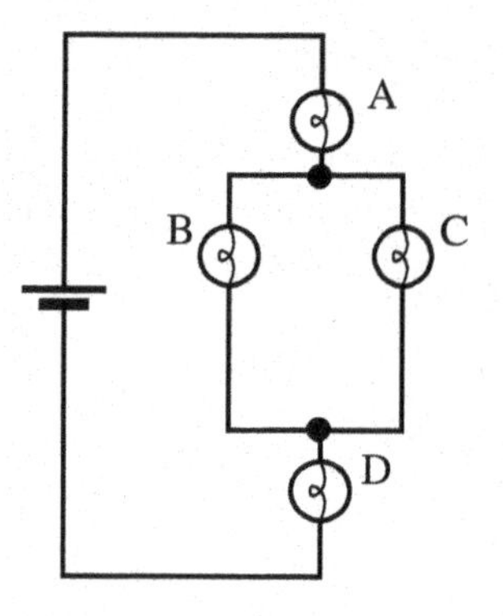

   a. Rank the bulbs in order from brightest to dimmest. If two bulbs have the same brightness, indicate that explicitly.

      Explain how you determined the ranking of the bulbs.

   b. Suppose that a switch has been added to the circuit as shown. The switch is initially closed.

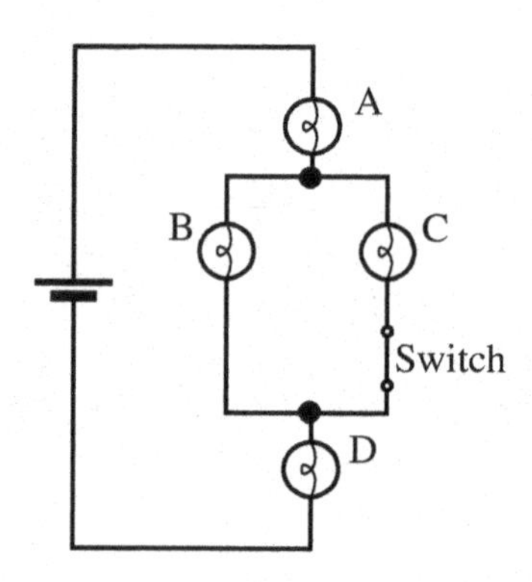

      When the switch is opened, will the current through bulb A *increase, decrease,* or *remain the same?* Explain how you determined your answer.

Name ____________________

1. The circuit at right consists of a bulb in series with an electrical "black box."

   The following are possible contents for the "black box." All the bulbs are identical. Box C consists of a single piece of connecting wire.

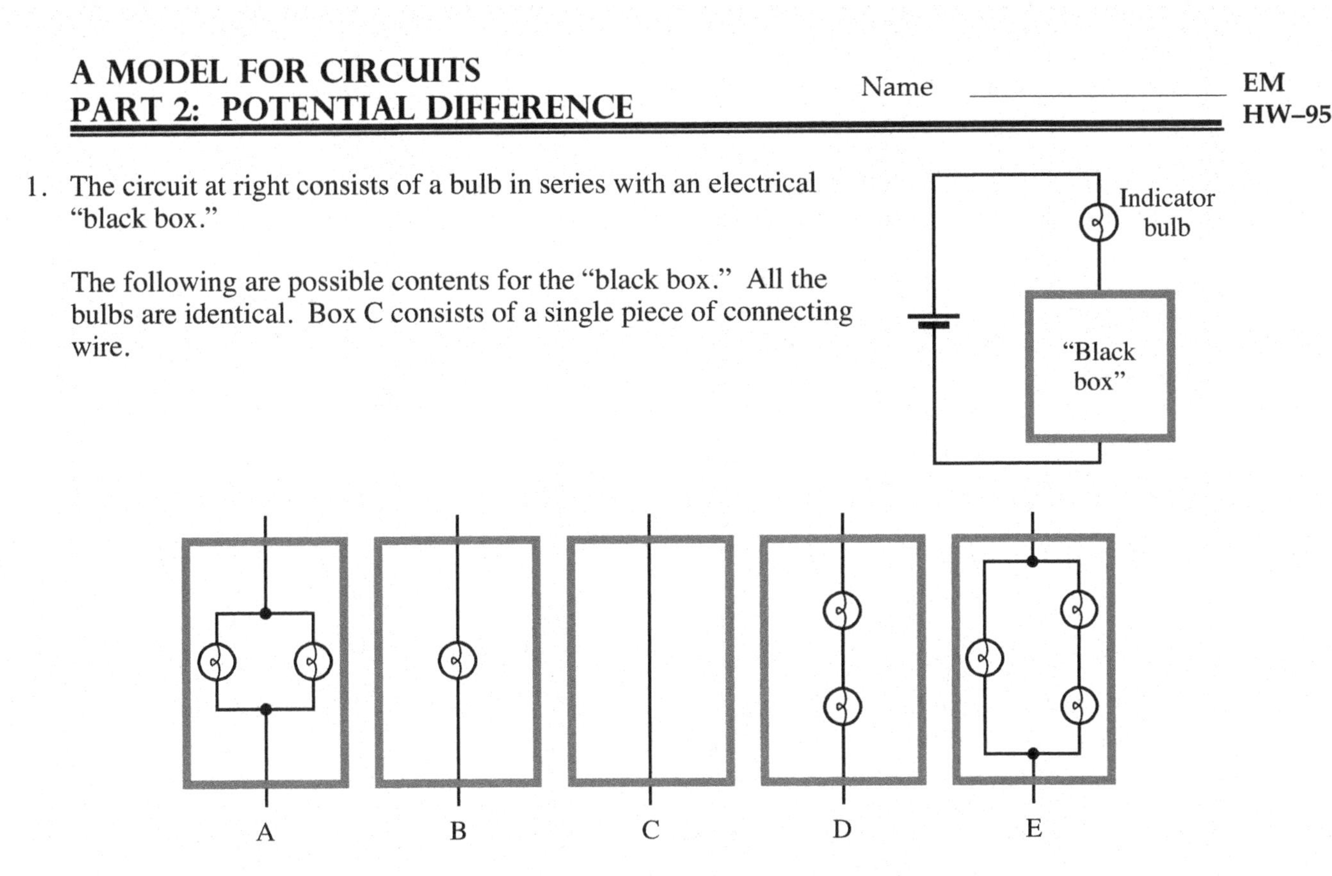

Suppose that each of the five boxes is placed, in turn, into the circuit. (Only one box is in the circuit at a time.)

a. Draw the appropriate five circuit diagrams below (show the *contents* of the boxes in place of the box itself).

b. Use the model that we have developed to rank the five circuits according to the brightness of the *indicator bulb* in those circuits.

c. Rank the boxes according to their equivalent resistance. Explain. (Do not calculate the resistance of each network; use the model instead.)

2. In this problem, box A and box B contain unknown combinations of light bulbs. Bulb 1 is identical to bulb 2. The batteries are ideal.

   a. In the circuit at right, the voltage across bulb 1 and the voltage across box A are equal. What, if anything, can you say about the resistance of box A compared to the resistance of bulb 1? Explain.

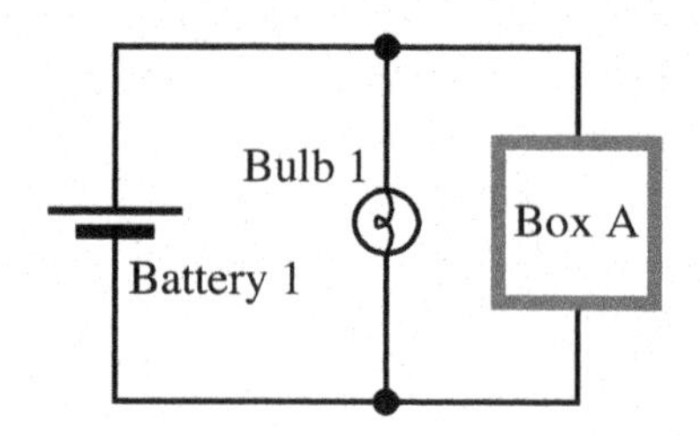

   Write an expression for the voltage across the battery (V_{Bat}) in terms of the voltage readings across box A and across the bulb ($V_{Box\ A}$, $V_{Bulb\ 1}$).

   b. In the circuit at right the voltage across bulb 2 and the voltage across box B were found to be equal. What, if anything, can you say about the resistance of box B compared to the resistance of bulb 2?

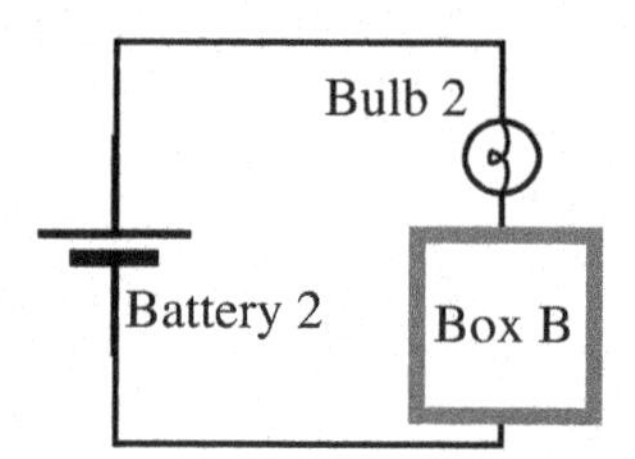

   Write an expression for the voltage across the battery (V_{Bat}) in terms of the voltage readings across box B and across the bulb ($V_{Box\ B}$, $V_{Bulb\ 2}$).

c. Box A and box B are now interchanged. It is observed that bulb 2 is now brighter than it was when box B was in that circuit.

   i. Is the resistance of box A *greater than, less than,* or *equal to* the resistance of box B? Explain.

   ii. Has the current through battery 1 changed? If so, how?

   iii. Has the current through battery 2 changed? If so, how?

3. The bulbs in the circuit at right are identical and the battery is ideal.

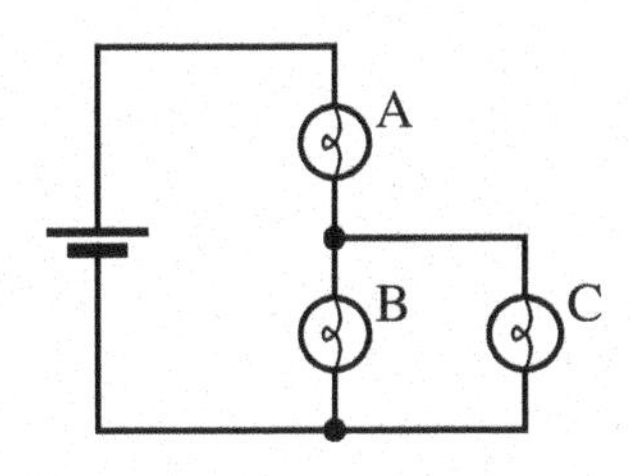

   a. Consider the circuit as shown.

      i. Rank bulbs A, B, and C in order of brightness. Explain how you determined your answer.

      ii. Rank the voltages across the bulbs. Explain.

      iii. Write an equation that relates the voltage across bulbs A and B to the battery voltage.

      iv. Is the voltage across bulb A *greater than, less than,* or *equal to* one half the battery voltage? Explain your reasoning.

b. A student cuts the wire between bulbs A and C as shown.

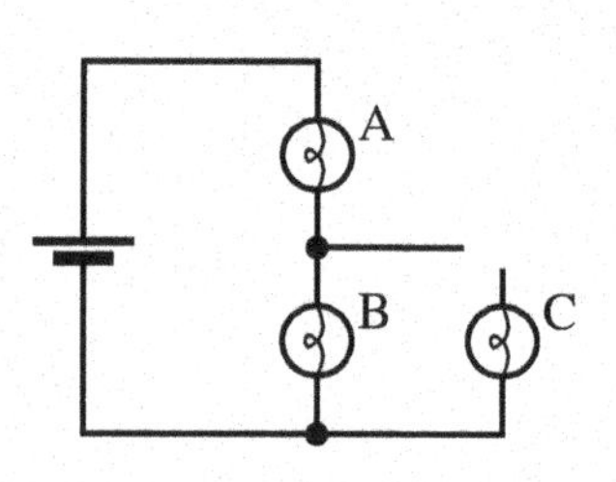

i. Rank bulbs A, B, and C in order of brightness. Explain how you determined your answer.

ii. Rank the voltages across the bulbs. Explain.

iii. Write an equation that relates the voltage across bulbs A and B to the battery voltage.

iv. Is the voltage across bulb A *greater than, less than,* or *equal to* one half the battery voltage? Explain your reasoning.

c. Consider the following discussion between two students regarding the change in the circuit when the wire in the circuit is cut (part b):

Student 1: *"I think that bulb B will get brighter. Bulb B used to share the current with bulb C, but now it gets all the current. So bulb B will get brighter."*

Student 2: *"I don't think so. Now there aren't as many paths for the current, so the resistance in the circuit has increased. Since the resistance in the circuit has gone up, the current in the circuit decreases. Bulb B will get dimmer."*

i. Is Student 1 correct? Why or why not?

ii. Is Student 2 correct? Why or why not?

iii. Use what you have learned about voltage to determine whether bulb B will *become brighter, become dimmer,* or *stay the same brightness* when the wire is cut. Explain how you determined your answer.

4. The bulbs in the circuit shown are identical. Treat the battery as ideal in answering all the questions.

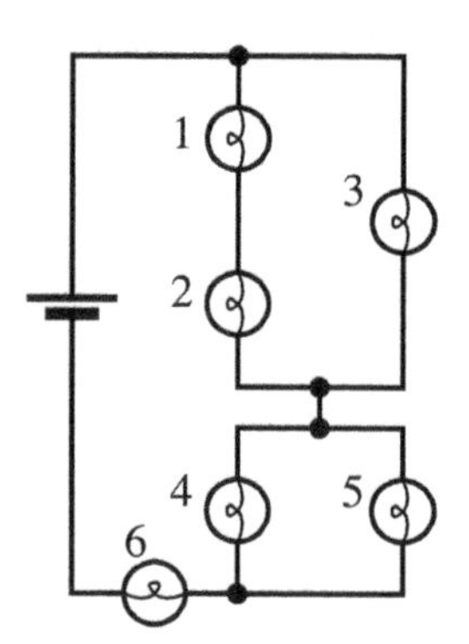

a. Rank bulbs 1–6 in order of brightness. Explain your reasoning.

b. Rank the voltages across the bulbs. Explain your reasoning.

c. Write an equation that relates the voltage across bulbs 3, 5, and 6 to the battery voltage.

d. Bulb 1 is removed from its socket.

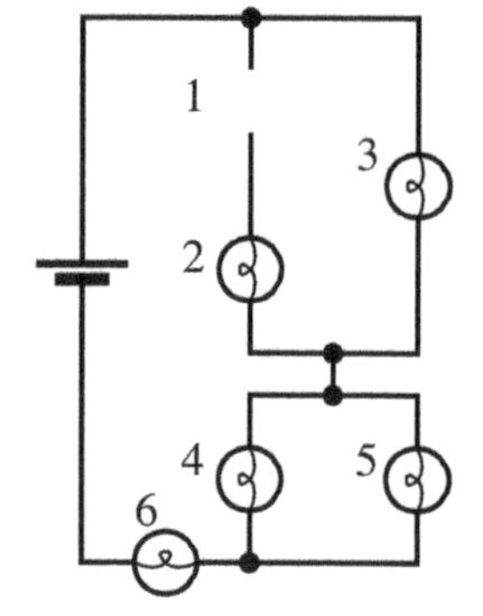

i. Does the brightness of bulb 2 *increase, decrease,* or *remain the same?* Explain.

ii. Does the brightness of bulb 6 *increase, decrease,* or *remain the same?* Explain.

iii. Does the brightness of bulb 3 *increase, decrease,* or *remain the same?* Explain.

1. The circuit at right contains a battery, a bulb, a switch, and a capacitor. The capacitor is initally uncharged.

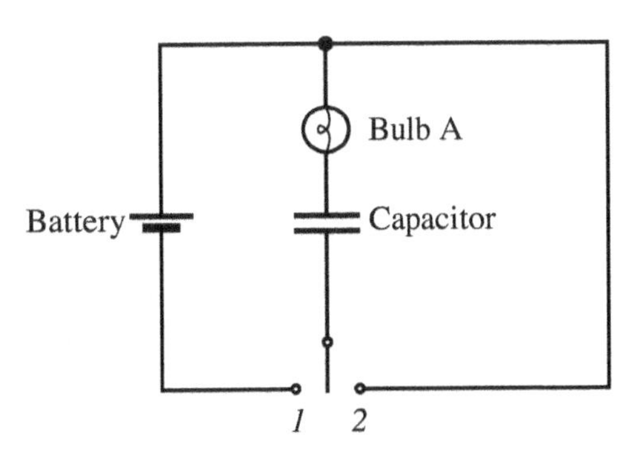

   a. Describe the behavior of the bulb in the two situations below.

      i. The switch is first moved to position *1*. Describe the behavior of the bulb *from just after the switch is closed until a long time later*. Explain.

      ii. The switch is now moved to position *2*. Describe the behavior of the bulb *from just after the switch is closed until a long time later*. Explain your reasoning.

   b. A second identical bulb is now added to the circuit as shown. The capacitor is discharged.

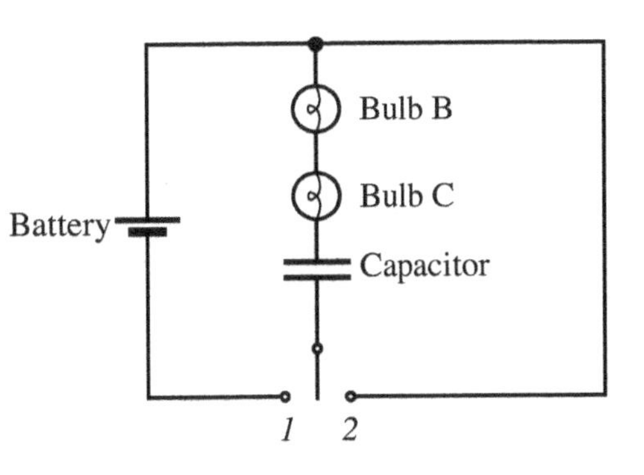

      i. The switch is now moved to position *1*.

      Describe the behavior of bulbs B and C *from just after the switch is closed until a long time later*. Explain.

      How does the initial brightness of bulb C compare to the initial brightness of bulb A in question i of part a? Explain your reasoning.

      A long time after the switch is closed, is the potential difference across the capacitor *greater than, less than,* or *equal to* the potential difference across the battery? Explain.

ii. The switch is now moved to position 2.

Describe the behavior of bulbs B and C *from just after the switch is closed until a long time later*. Explain your reasoning.

How does the initial brightness of bulb C compare to the initial brightness of bulb A in question ii of part a? Explain your reasoning.

A long time after the switch is closed, is the potential difference across the capacitor *greater than, less than,* or *equal to* the potential difference across the battery? Explain.

2. Two identical bulbs and a capacitor are connected to an ideal battery as shown. The capacitor is initially uncharged.

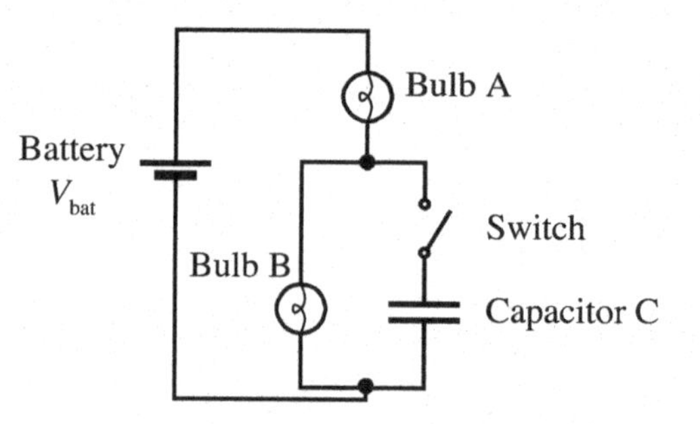

a. Just after the switch is closed:

- what is the potential difference across bulb A, across bulb B, across the capacitor C, and across the battery? Explain.

- rank the currents i_A, i_B, i_C, and i_{bat}. Explain your reasoning.

b. A long time after the switch is closed:

- rank the currents i_A, i_B, i_C, and i_{bat}. Explain your reasoning.

- what is the potential difference across bulb A, bulb B, the capacitor C, and the battery? Explain.

c. Summarize your results by describing the behavior of bulb A and of bulb B *from just after the switch is closed until a long time later*.

1. Answer the questions below on the basis of your experience with charges and magnets.

   a. Based on your experience with *electric field lines:*

      - how should the *direction* of the magnetic field at every point be related to the *magnetic field lines?*

      - how should the *strength* of the magnetic field at every point be reflected in the *magnetic field lines?*

   b. Carefully draw the *magnetic field lines* for the bar magnet shown below. Be sure to draw the field lines so that they include information about the *strength* and *direction* of the field both inside and outside the magnet.

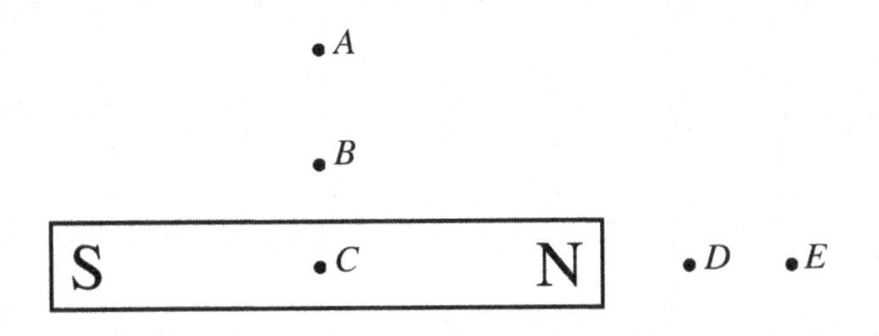

   c. Based on the magnetic field lines you have drawn, rank the magnitude of the magnetic field at points *A*–*E*.

2. Two identical magnets are placed as shown. Using different colored pens sketch the approximate magnetic field vectors at the four labeled points for:

   - just the horizontal top magnet,
   - just the vertical bottom magnet, and
   - when both are present.

   Explain how you determined the field vectors for the case when both magnets are present.

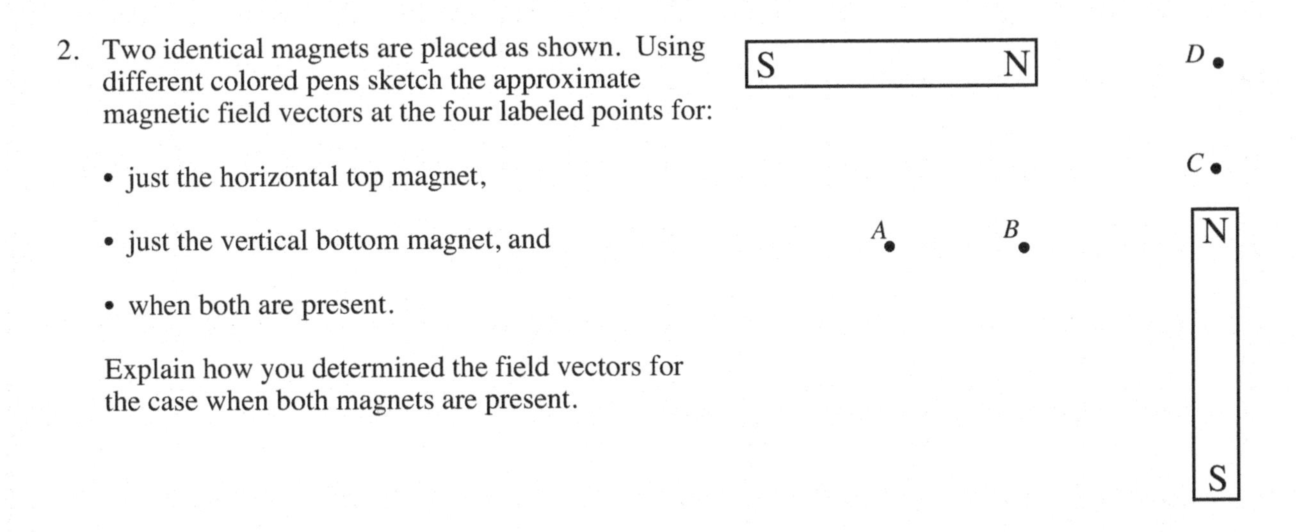

3. Three metal bars, labeled 1, 2, and 3, are marked A and B on either end as shown.

   The following observations are made by a student:

   - end 1B repels end 3A
   - end 1A attracts end 2B
   - end 2B attracts end 3B

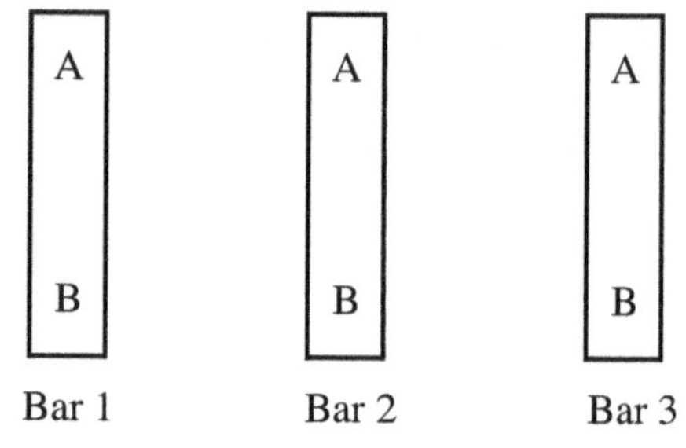

   a. To which of your three classes from section I of the tutorial *Magnets and magnetic fields* could bar 1 belong? Explain your reasoning and the characteristics that define each of your classes.

   b. To which of your three classes could bar 2 belong? Explain your reasoning.

   c. Would end 2A *attract, repel,* or *neither attract nor repel* end 3A if the two ends were brought near one another? If it is not possible to tell for certain, what are the possibilities? Explain.

1. A magnet is hung by a string and then placed near a wire as shown. When the switch is closed, the magnet rotates such that the ends of the magnet move as indicated by the arrows.

   At the instant the switch is closed determine:

   - the direction of the current through the wire segment nearest the magnet. Explain.

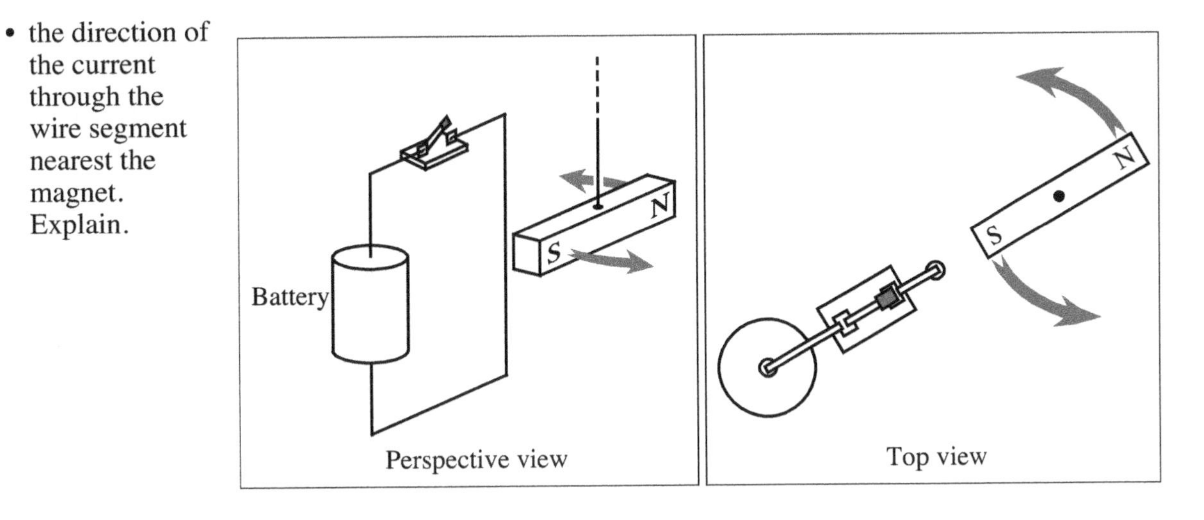

Perspective view Top view

   - the direction of the net force exerted by the magnet on the wire segment at the instant that the magnet is in the position shown. Explain.

2. Shown at right is a cross-sectional view of two long straight wires that are parallel to one another. One wire carries a current i_o out of the page, the other carries an equal current i_o into the page.

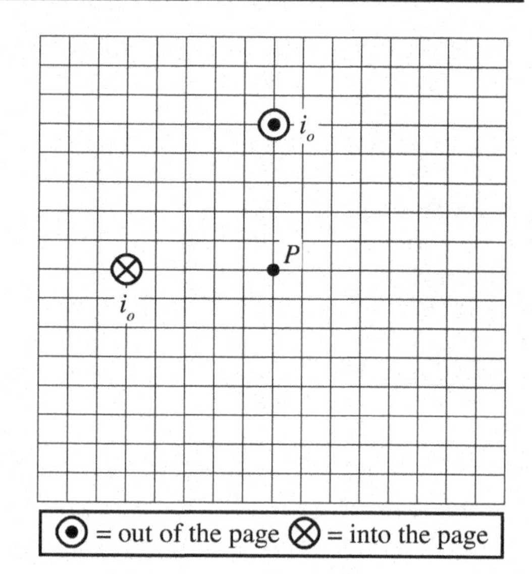

a. Draw a vector on the diagram to show the direction of the magnetic field, if any, at point *P*. Explain your reasoning.

b. Suppose that a third wire, carrying another current i_o out of the page, passes through point *P*.

Draw a vector on the diagram to indicate the magnetic force, if any, exerted on the current in the new wire at *P*. If the magnitude of the force is zero, indicate that explicitly. Explain your reasoning.

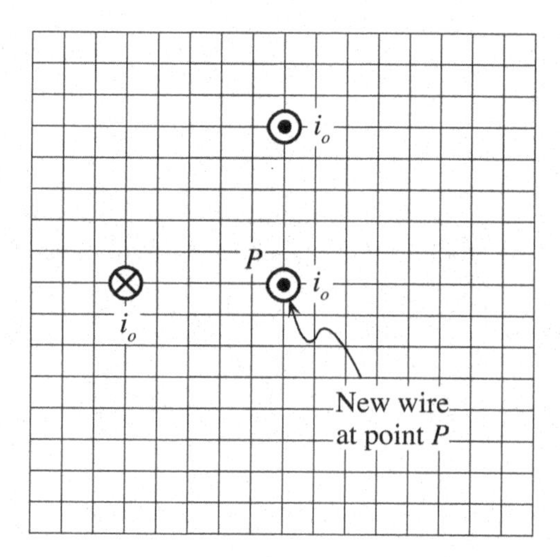

c. Suppose instead that the third wire (carrying the same current i_o out of the page) is placed such that the magnetic field at point *P* has zero magnitude. Determine the location of the third wire. (*Hint:* You will need to know how the magnetic field depends on the distance from the wire. This relationship can be found in your text.)

Clearly indicate on the diagram at right the correct location of the new wire. Explain how you determined your answer.

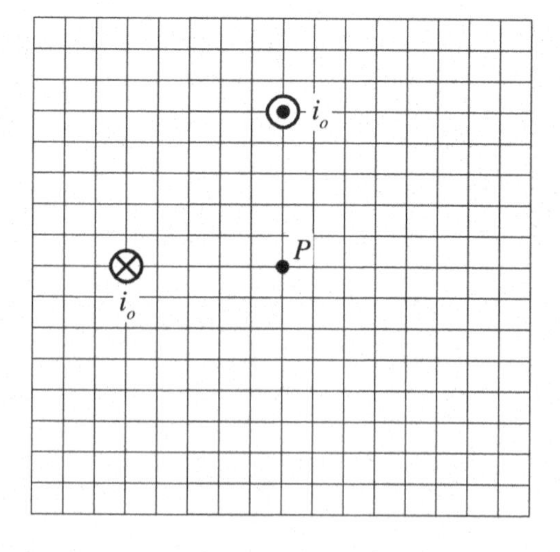

1. The diagram at right shows a copper wire *loop* held in place near a *solenoid*. The switch in the circuit containing the solenoid is initially open.

   a. Use Lenz' law to predict whether current will flow through the wire of the loop in each of the following cases. Explain your reasoning.

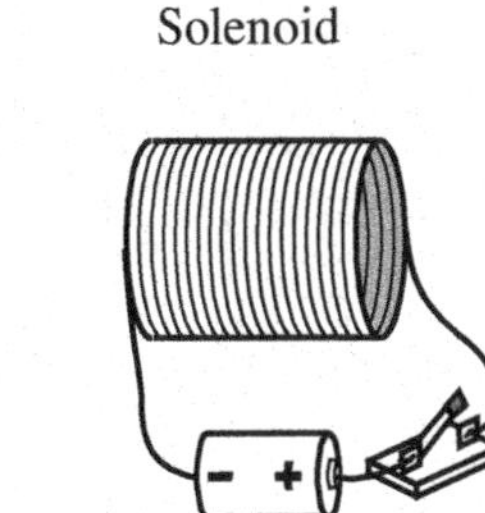

   - just after the switch has been closed

   - a long time after the switch has been closed

   - just after the switch has been reopened

   - a long time after the switch has been reopened

   b. For each of the cases in which you predicted that there will be an induced current, draw a diagram that illustrates:

   - the direction of the current through the wire of the loop,
   - the direction of the magnetic moment of the loop, and
   - the direction of the magnetic force exerted on the loop.

   Is the force on the wire loop in a direction that would tend to *increase* or *decrease* the change in net flux through the wire loop?

2. A copper wire loop is constructed so that its radius, r, can change. It is held near a solenoid that has a constant current through it.

   a. Suppose that the radius of the loop were increasing. Use Lenz' law to explain why there would be an induced current through the wire. Indicate the direction of that current.

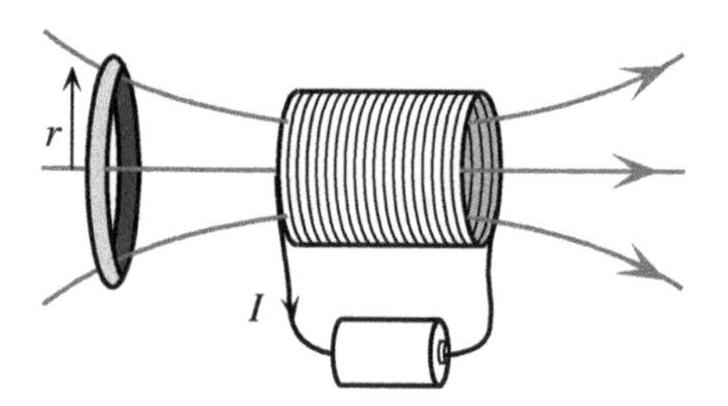

   b. Check your answer regarding the direction of the induced current by considering the magnetic force that is exerted on the charge in the wire of the loop.

   c. Find:

   - the direction of the magnetic moment of the loop and

   - the direction of the force exerted on the loop by the solenoid.

3 A copper wire loop is initially at rest in a uniform magnetic field. Between times $t = t_o$ and $t = t_o + \Delta t$ the loop is rotated about a vertical axis as shown.

Will current flow through the wire of the loop during this time interval? If so, indicate the direction of the induced current and explain your reasoning. If not, explain why not.

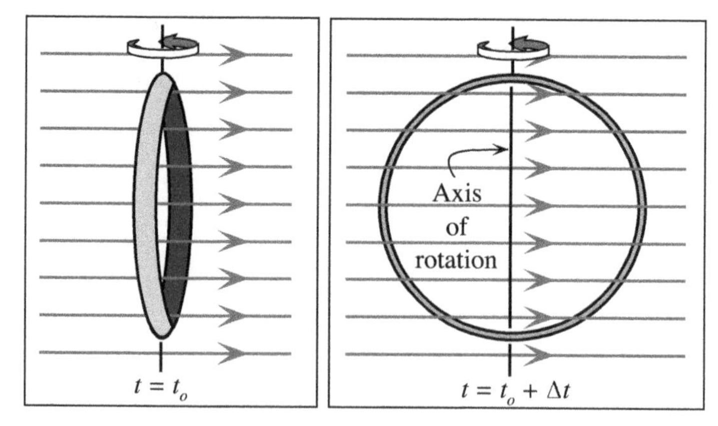

1. A metal loop is attached to an axle with a handle as shown. The north pole of a magnet is placed below the loop and the handle turned so that the loop rotates counterclockwise at a constant angular speed ω.

   a. On the two diagrams below, indicate the direction of the induced current in the loop at each of the instants shown. If the current is zero, state that explicitly. Explain how you determined your answers.

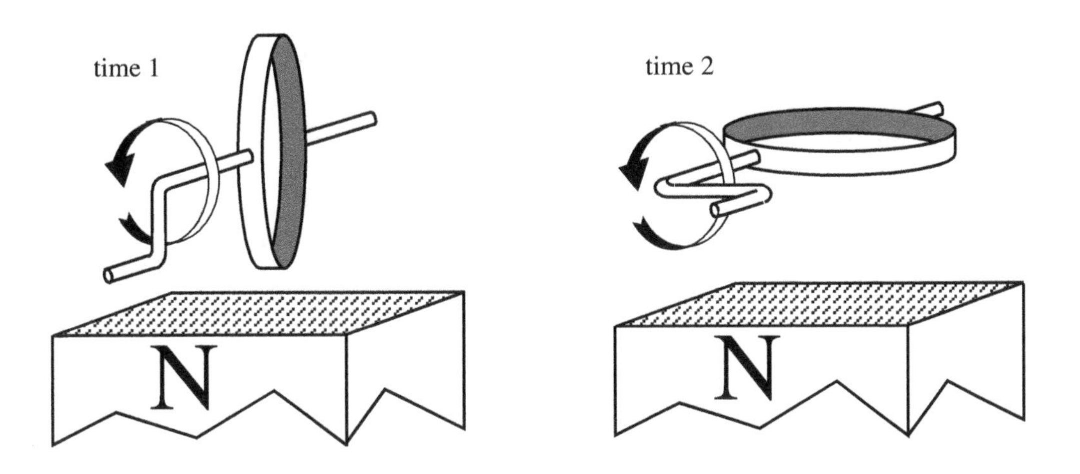

   b. Suppose the loop were replaced by a second loop that is identical to the first except for a small cut in it (as shown). The loop is rotated as before.

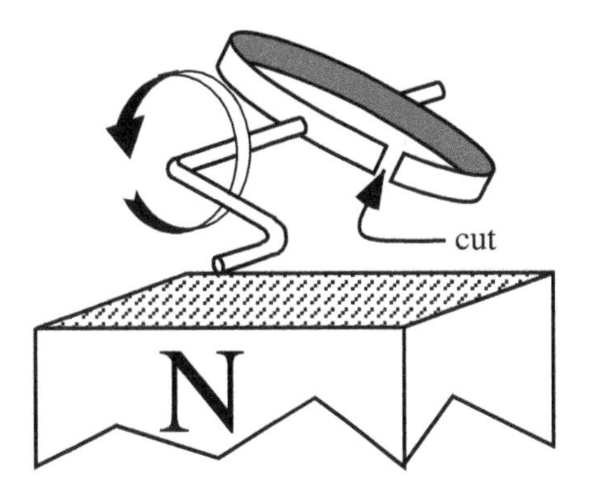

      i. How does the maximum induced emf in the uncut loop compare to the maximum induced emf in the cut loop? Explain.

      ii. How does the maximum induced current in the uncut loop compare to the maximum induced current in the cut loop? Explain your reasoning.

2. Five loops are formed of copper wire of the same gauge (cross-sectional area). Loops 1–4 are identical; loop 5 has the same height as the others but is longer. At the instant shown, all the loops are moving at the same speed in the directions indicated.

   There is a uniform magnetic field pointing out of the page in region I; in region II there is no magnetic field. Ignore any interactions between the loops.

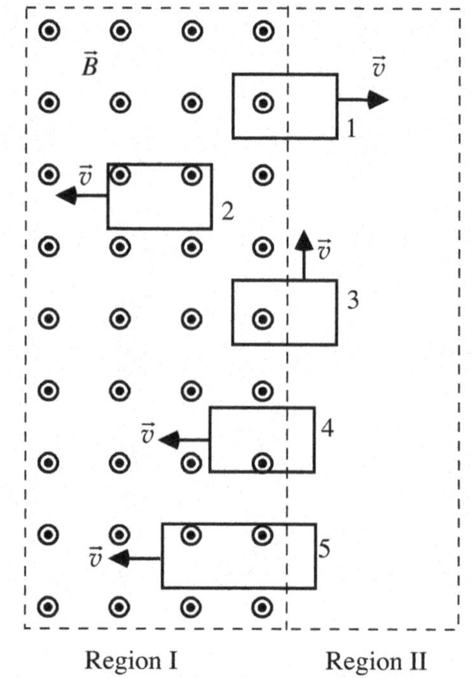

   a. For any loop that has an induced current, indicate the direction of that current.

   b. Rank the magnitudes of the *emfs* around the loops. Explain your reasoning.

   c. Rank the magnitudes of the currents in the loops. Explain your reasoning.

Waves

1. Two pulses on opposite sides of a spring are moving toward each other. The diagrams below show the pulse locations at four successive instants.

   On each diagram, sketch the shape of the spring for the instant shown.

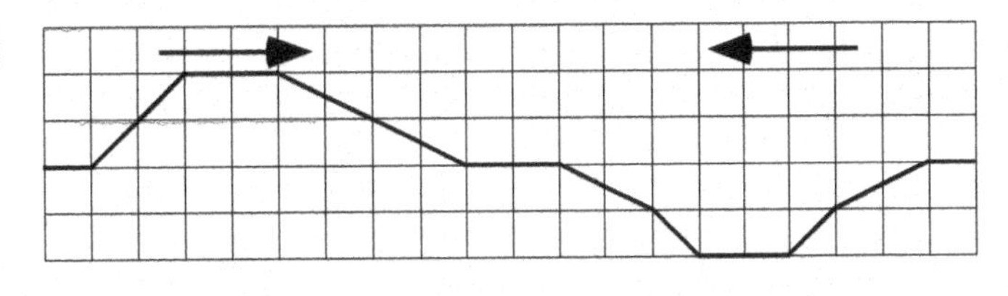

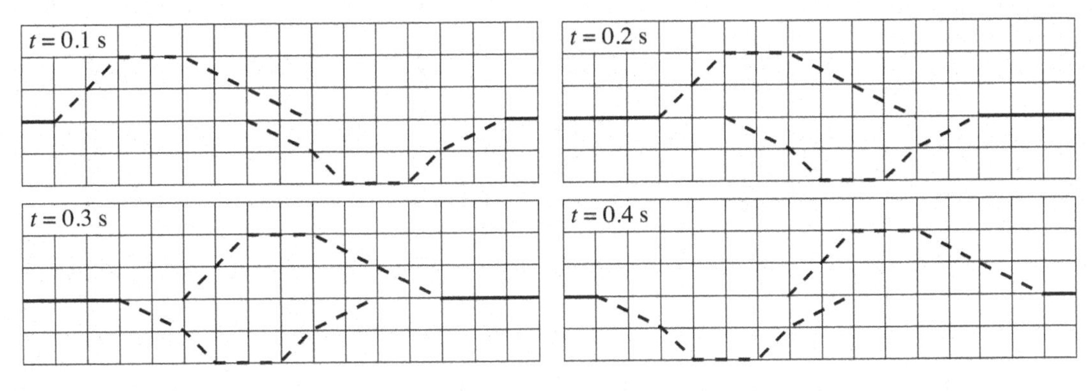

2. A pulse is moving toward the end of a spring that is attached to a wall. The diagram at right shows the pulse at $t = 0.0$ s. The leading edge of the pulse reaches the wall at $t = 0.4$ s.

   a. In the spaces provided below right, carefully draw the shape of the spring at $t = 0.8$ s and $t = 1.2$ s.

      (*Note:* The pulse is in the process of being reflected at $t = 0.8$ s.)

   b. In the space below, explain the steps that you followed in part a in order to determine the shape of the spring.

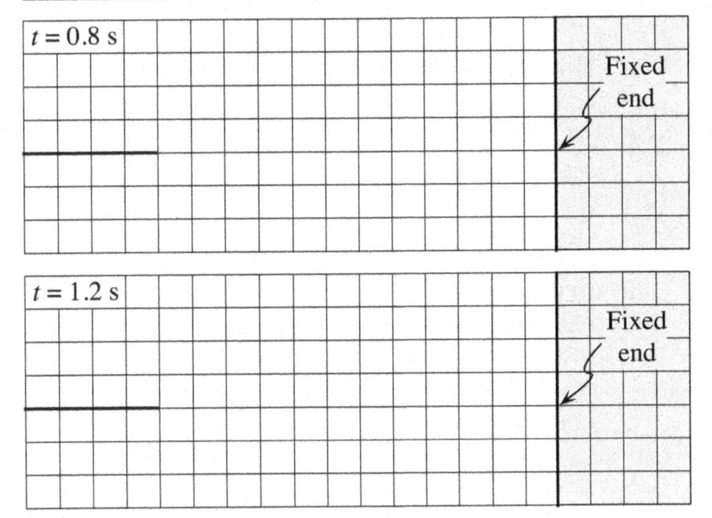

3. In this problem, you will determine and apply the appropriate boundary condition for reflection from a free-end of a spring (*i.e.*, an end that is free to move in a direction transverse to the length of the spring).

   a. We begin by considering the forces exerted on a ring that is connected to a spring and that is free to slide along a rod. (See the top view diagram at right.)

   Assume that the ring is *massless* and that the rod is *frictionless*.

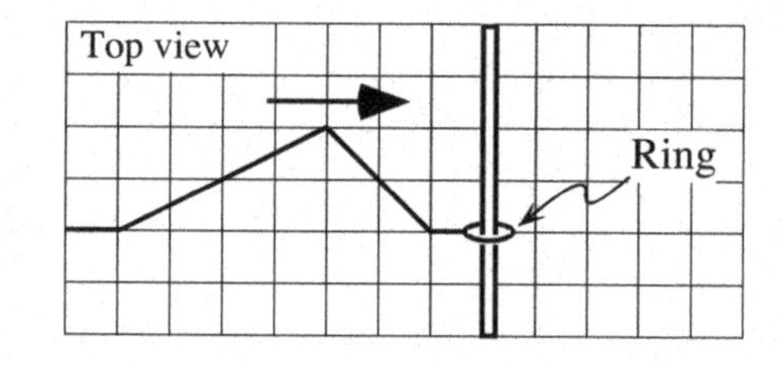

   i. What is the *net force* on the ring? (*Hint:* Consider what happens to the net force on an object as its mass approaches zero.) Explain.

   Does the net force on a massless ring depend on the acceleration of the ring? Explain.

   What is the *magnitude* of the gravitational force exerted on the ring? (Recall that the ring is massless.)

   ii. Does the force exerted on the ring by the rod have a component that is *parallel to the rod?* Explain. (*Hint:* Recall the assumptions made above.)

   iii. In the space at right, draw and label a free-body diagram for the ring at the instant when it is farthest from its equilibrium position. (*Hint:* Which objects are in *contact* with the ring?)

   Free-body diagram for massless ring

   Check that your free-body diagram is consistent with your answers to parts i and ii.

iv. For the instant shown in your free-body diagram, does the force exerted on the ring by the spring have a component that is *parallel to the rod?* Explain.

Would your answer above differ if you considered an instant when the free end was *not* at its farthest point from its equilibrium position? Explain.

v. What do your results in part iv suggest about the shape of the spring very near the ring? In particular, which of the diagrams at right best represents the "slope" of the spring at the point where the spring is connected to the ring?

(The correct answer to part v is the boundary condition for a free-end reflection.)

b. At the end of the tutorial *Superposition and reflection of pulses,* you observed a demonstration of a pulse reflected from the free end of a spring. A row of paper cups was placed near a spring as shown in the top view diagram below. A pulse with an amplitude slightly less than the cup-spring distance was sent down the spring.

On the diagram, indicate which cup(s) were hit by the spring during the demonstration.

On the basis of your observation, did the free end of the spring have a maximum displacement that was *greater than, less than,* or *equal to* the amplitude of the incident pulse?

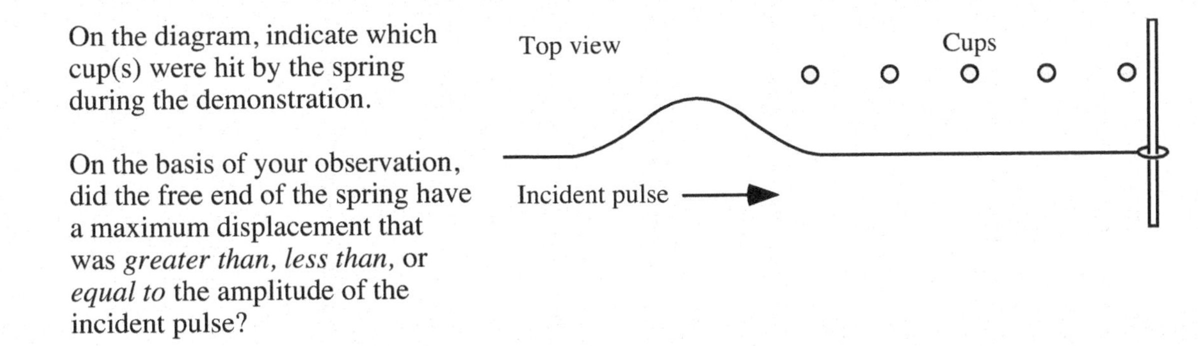

On the basis of your answer above, were the incident and reflected pulses on the *same side* of the spring or on *opposite sides* of the spring? Explain.

In the tutorial *Superposition and reflection of pulses,* we developed a model that we can use to predict the shape and orientation of a pulse reflected from the fixed end of a spring. We can develop a similar model for free-end reflection. In this case as well, we imagine that the spring extends past the free end. We then imagine sending a pulse with the appropriate shape and location on this imaginary portion of spring toward the incident pulse so that, as the pulses pass each other, the appropriate boundary condition at the free-end is satisfied.

c. Consider a pulse incident on the free end of a spring as shown.

   i. Would the reflected and incident pulses be on the *same* side of the spring or on *opposite* sides?

      Would you expect the incident and reflected pulses to have the *same* leading edge or *different* leading edges? Explain.

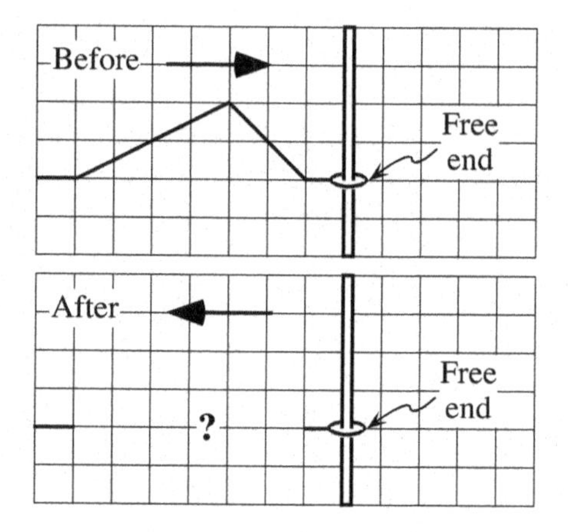

   ii. Sketch your prediction for the shape of the reflected pulse on the diagram at right.

d. At $t = 0.0$ s a pulse with speed 1.0 m/s is incident on the free end of a spring as shown.

   i. Predict the shape of the spring at $t = 0.2$ s, 0.4 s, and 0.6 s.

$t = 0.0$ s (1 square = 10 cm) Free end

$t = 0.2$ s (1 square = 10 cm)

$t = 0.4$ s (1 square = 10 cm)

$t = 0.6$ s (1 square = 10 cm)

   ii. Are your diagrams consistent with the boundary condition that you determined in part a for a free end of a spring? If so, explain how you can tell. If not, check your answers to part c and resolve any inconsistencies.

In the tutorial *Superposition and reflection of pulses,* we illustrated transverse pulses using idealized shapes that have sharp corners or "kinks." While this approximation is convenient for applying the principle of superposition, it can lead to inconsistencies when considering reflection from a free end of a spring. This inconsistency is addressed in the next problem.

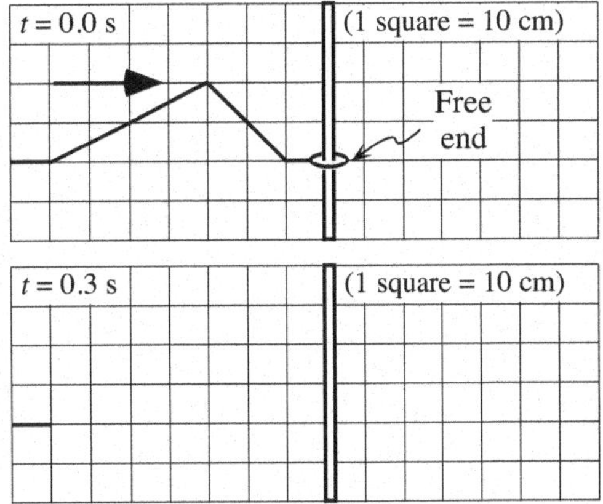

e. Consider again the situation from part d, in which a pulse with a speed of 1.0 m/s is incident on the free end of a spring.

   i. In the space provided, show that the shape of the spring at $t = 0.3$ s (determined by using your model for free-end reflection) is *not* consistent with the correct boundary condition for this case.

   ii. Note that the peak of the pulse is drawn as a sharp "kink" in the spring. How does this "kink" lead to the inconsistency mentioned above?

4. The diagram at right shows two pulses. One is being reflected from a fixed end; the other, from a free end.

   The diagram has several errors. Describe each error that you can find.

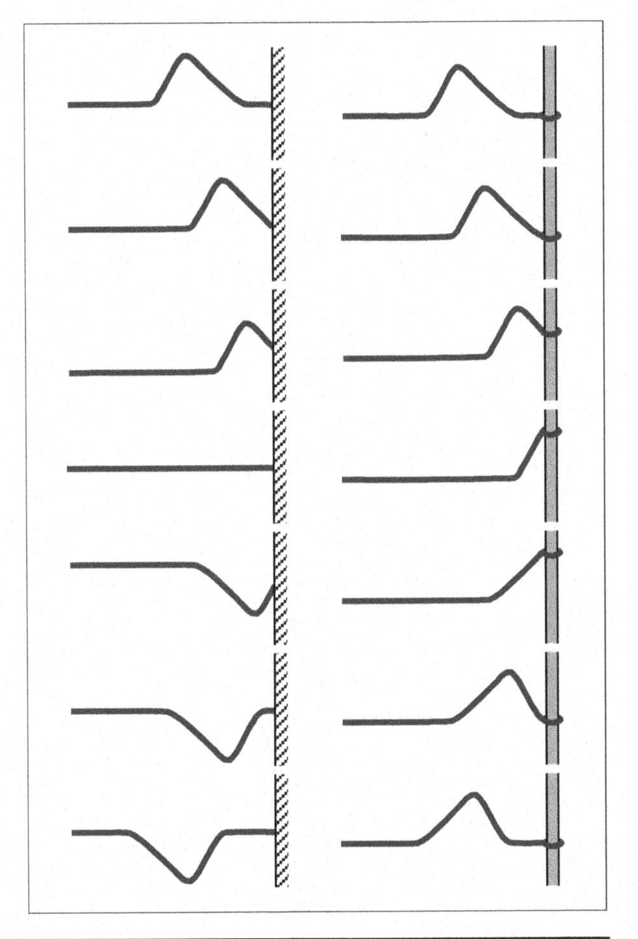

1. A pulse in spring L is moving to the right as shown in the first diagram below. A short time later, a reflected pulse and a transmitted pulse will travel away from the junction and toward the walls. The transverse displacements of the springs have been exaggerated for clarity.

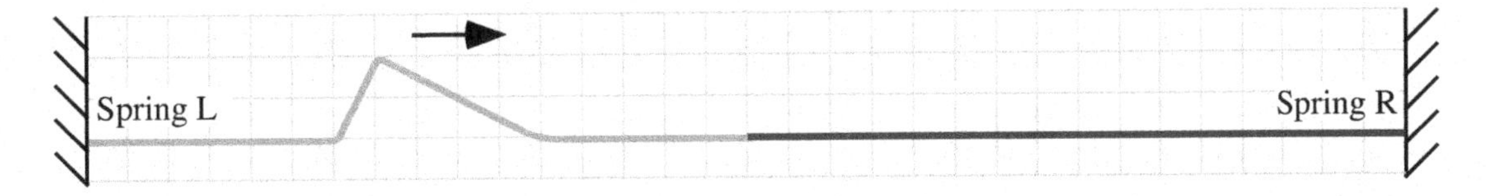

Sketch the shape of the springs at an instant before the transmitted and reflected pulses reach the walls in the following cases: (1) the wave speed in spring R is *less than* the wave speed in spring L, and (2) the wave speed in spring R is *greater than* the wave speed in spring L. Your drawings should be *qualitatively* correct; however, you are not expected to show the correct relative amplitudes of the pulses.

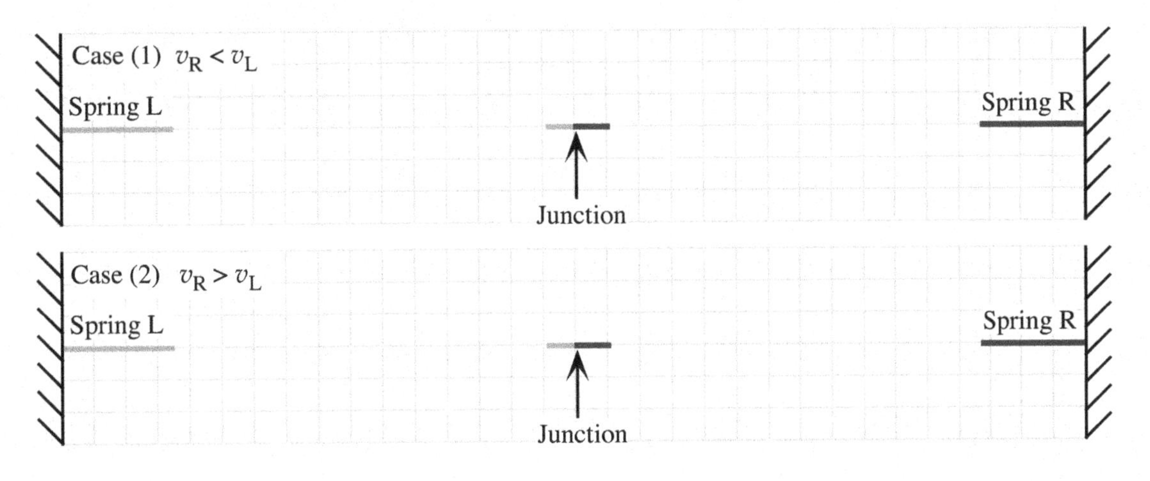

2. The diagram below represents a snapshot of two springs at an instant *after* a pulse has reached the boundary between them.

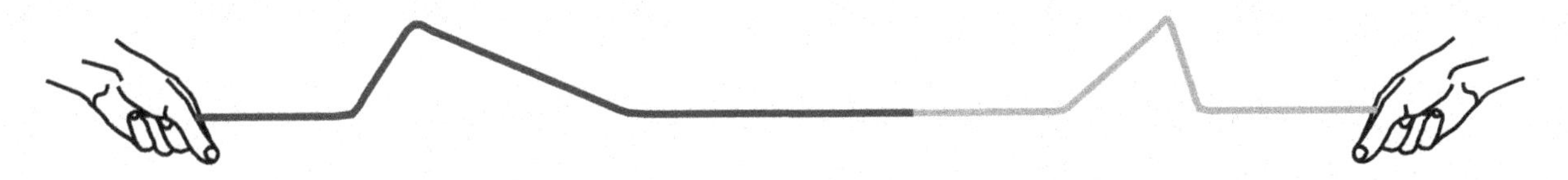

a. On the diagram, clearly label which spring has the *larger* wave speed. Explain how you could tell from the diagram.

b. Is the reflection at the boundary between the springs more like reflection from a fixed end or a free end? Explain how you can tell from the diagram.

c. On the diagram, label which pulse is the *reflected* pulse and which is the *transmitted* pulse. Explain your reasoning.

3. Each of the diagrams below represents a snapshot of two springs at an instant *after* a pulse has reached the boundary between them. The linear mass density, μ, is greater for the tightly coiled spring than for the other spring.

   Each diagram *may* contain flaws that would not be observed with real springs. For each diagram:

   a. Determine whether there is a flaw. If there is a flaw, describe it (a single flaw is sufficient) and continue to the next diagram. If there are no flaws, answer parts b and c.

   b. Make a sketch that shows the shape, width, and direction of motion of the incident pulse.

   c. Determine which pulse is the reflected pulse and which is the transmitted pulse.

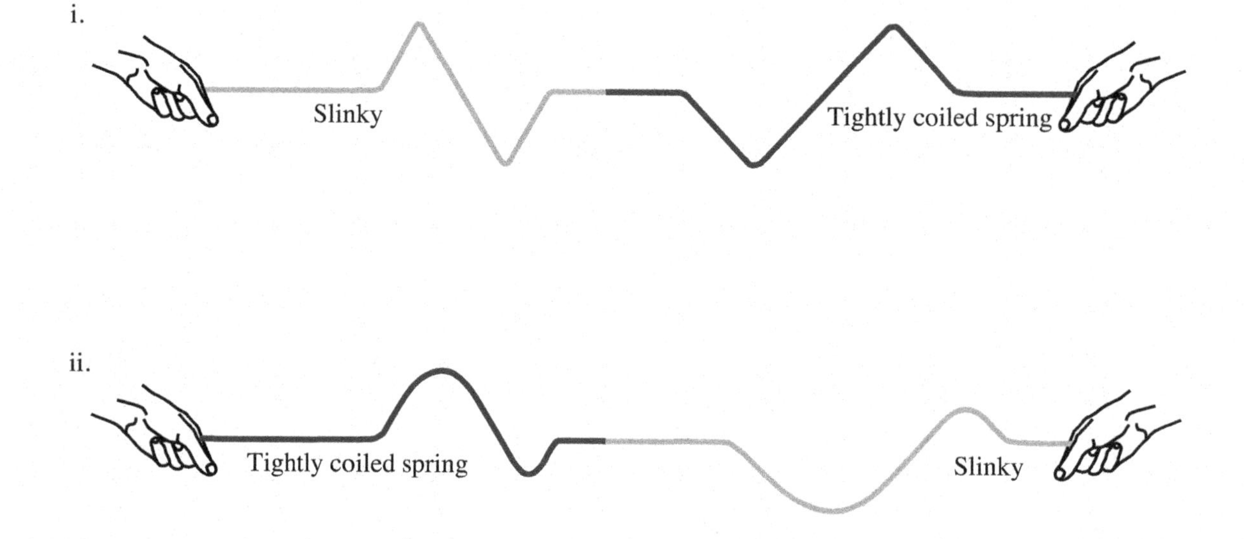

4. The figure at right has several errors. How many can you find? Explain briefly.

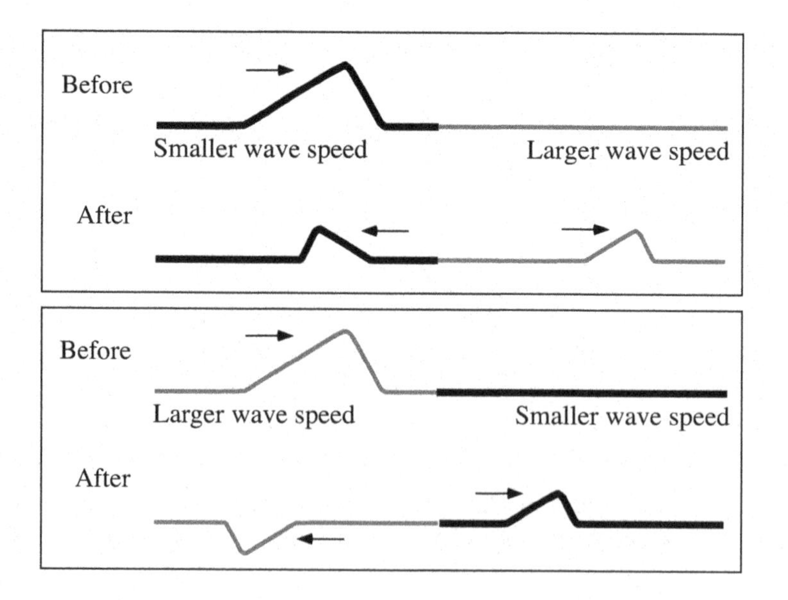

1. a. The terms below are often used to characterize periodic waves. Give a physical interpretation of each term. (For example, a physical interpretation of *speed,* in the case of uniform motion of an object, might be, "the distance the object moves in one unit of time." "How fast the object is moving" would *not* be an acceptable interpretation.)

   - period *(T)*

   - frequency *(f* or *ν)* (*Note:* "The reciprocal of the period" or "$1/T$" is *not* an interpretation.)

   - wavelength *(λ)*

   b. Show that the equation for the wave speed, $v = \lambda f$, comes directly from the definition of speed, in the case of uniform motion. (*Hint:* Recall the interpretation of speed from part a.)

   c. Explain why *T*, *λ*, and *f* (or *ν*) should *not* be applied to a *pulse*. (*Hint:* How is the interpretation of λ different from the width of a pulse?)

   d. For each of the periodic functions below, indicate the wavelength on the diagram.

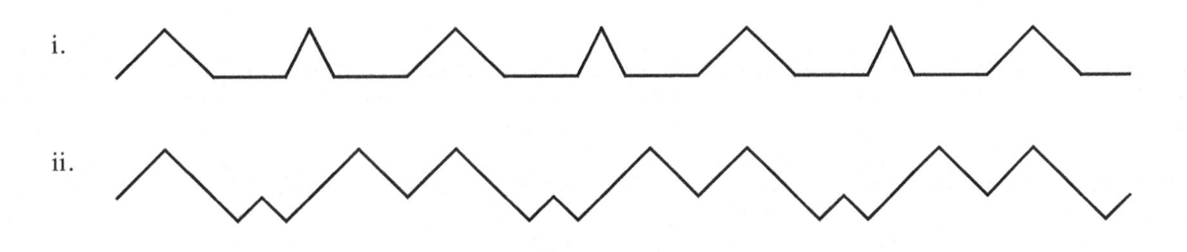

2. Determine whether each of the ray diagrams below has a flaw.

   If a diagram *has a flaw,* clearly describe how the physical situation is not consistent with observations that you made in tutorial (*e.g.*, a crest is transmitted as a crest).

   If a diagram *has no flaw:*

   - Use a straightedge to draw incident and transmitted wavefronts that are consistent with the rays and boundary shown.
   - If possible, determine in which medium the waves travel more quickly. If it is not possible to make this comparison, explain why not.

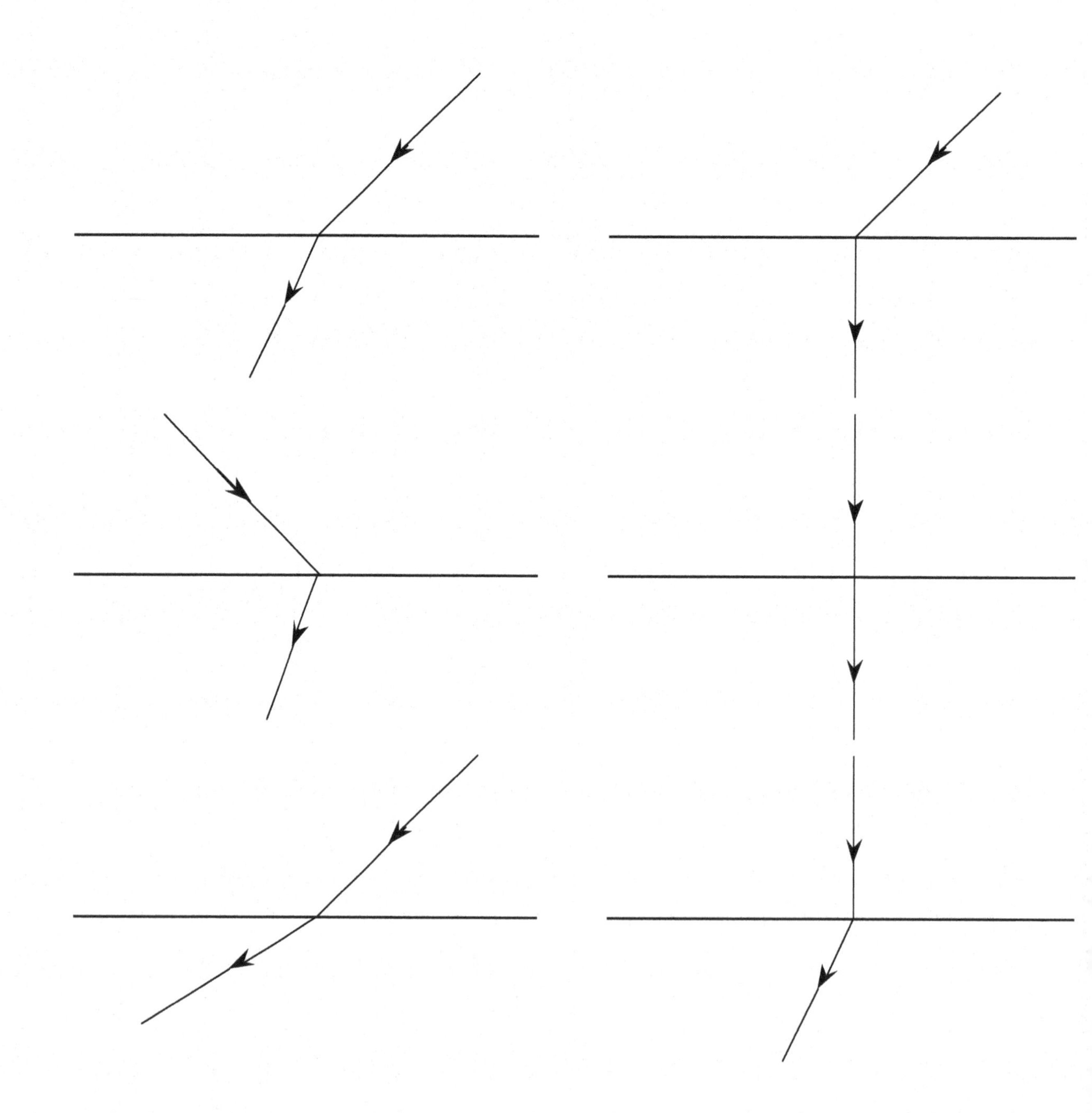

3. The diagram at right illustrates refraction of a wave as it propagates from a medium of larger wave speed to a medium of smaller wave speed.

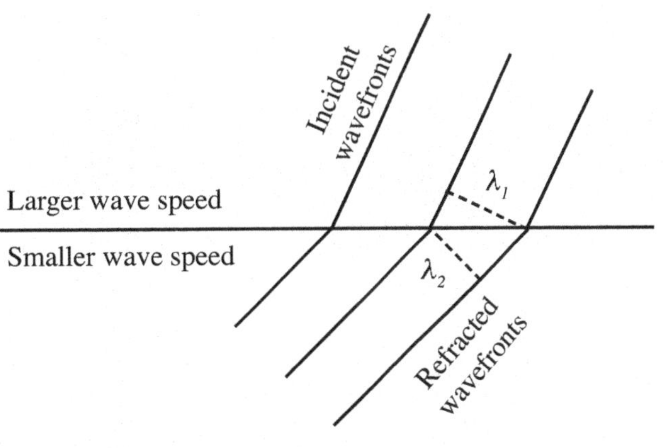

a. Use trigonometry to determine the mathematical relationship between the angle of incidence (θ_1), the angle of refraction (θ_2), the wavelength of the incident wave (λ_1), and the wavelength of the refracted wave (λ_2). Show your work.

b. Starting from the equation that you wrote above, derive a mathematical formula relating θ_1, θ_2, v_1, and v_2, where v_1 and v_2 are the speeds of the incident and refracted wave, respectively. Show your work.

c. Suppose the speed of the refracted wave were half that of the incident wave. Determine the angle of refraction for the following angles of incidence: 10°, 20°, 40°, and 80°.

Does the angle of refraction double when the angle of incidence doubles?

d. Would the relationship that you developed in parts a and b also apply to a wave passing from a medium of smaller wave speed to a medium of larger wave speed? Explain why or why not.

1. A long, thin steel wire is cut in half, and each half is connected to a different terminal of a light bulb. An electromagnetic (EM) plane wave ($\vec{E}(x, y, z, t) = E_o \sin(kx - \omega t)\,\hat{y}$, $\vec{B}(x, y, z, t) = B_o \sin(kx - \omega t)\,\hat{z}$) moves past the wire, as shown.

   a. In what direction is the wave propagating? Explain your reasoning.

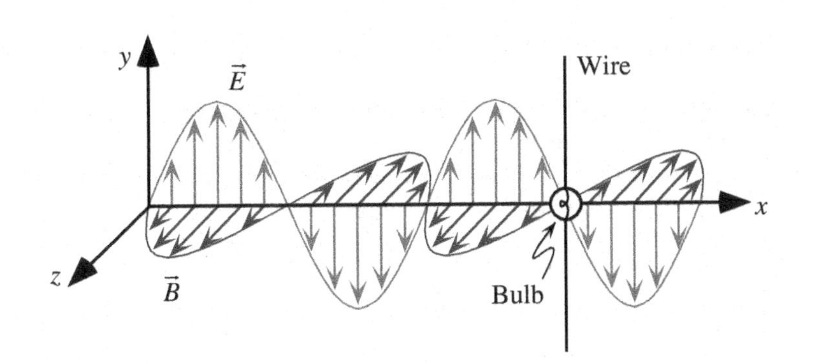

   b. Suppose the wire were oriented parallel to the *y*-axis, as shown above.

   Would the bulb glow in this case? Explain.

   c. Suppose instead that the wire were positioned as described below. Would the brightness of the bulb be *greater than, less than,* or *equal to* the brightness that it had in part b? Explain your reasoning in each case.

      i. The wire is parallel to the *y*-axis but with its bottom end located on the *x*-axis (*i.e.*, the wire is shifted upward a distance equal to half its length).

      ii. The wire is tilted so that it makes an angle of 40° with respect to the *y*-axis but is still parallel to the *y*–*z* plane. (See diagram at right.)

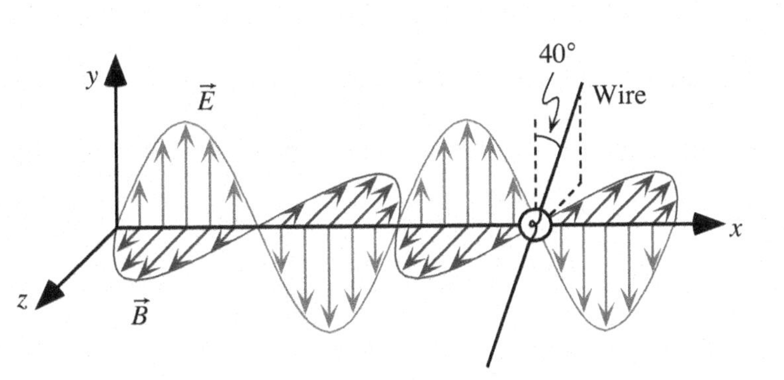

2. A radio-frequency EM plane wave (not shown) propagates in the +z-direction. A student finds that her portable radio obtains the best reception of the wave when the antenna is parallel to the x–y plane making an angle of 60° with respect to the y-axis. (See the diagram below.)

    a. Consider an instant when the fields are non-zero at the location of the antenna.

    On the diagram at right, draw and label arrows to indicate (1) the direction of the electric field and (2) the direction of the magnetic field. Explain your reasoning. (*Note:* More than one answer is possible.)

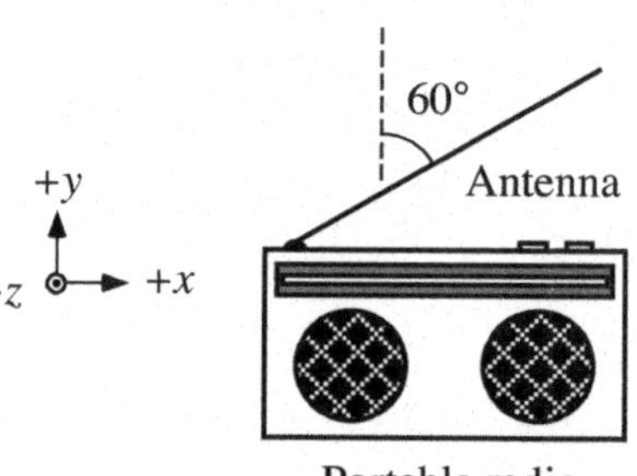

    b. How would your answers to part a be different if the wave were propagating in the −z-direction instead of the +z-direction? (*Note:* More than one answer is possible.) Explain.

3. In a region that does not contain current-carrying wires, the magnetic field is found to be $\vec{B}(x, y, z, t) = B_o \sin(kx - \omega t)\,\hat{z}$.

    Show that in such a region it is not be possible to have an electric field that is equal to *zero* for all x and t. (*Hint:* Set $i_{\text{encl}} = 0$ in Ampère's law:

$$\oint \vec{B} \cdot d\vec{l} = \mu_o i_{\text{encl}} + \mu_o \varepsilon_o \frac{d\Phi_E}{dt}$$

and consider the quantity $\oint \vec{B} \cdot d\vec{l}$ evaluated around the loop $5 \rightarrow 6 \rightarrow 7 \rightarrow 8 \rightarrow 5$ in the x–z plane, shown below.)

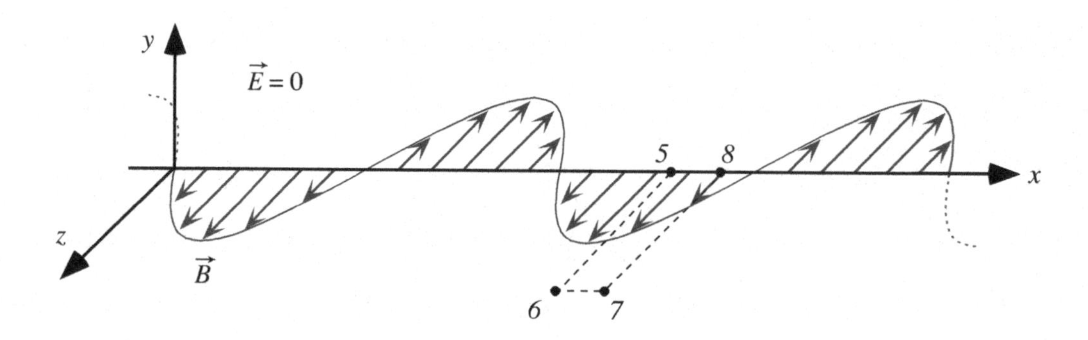

Optics

1. A mask containing a hole in the shape of the letter *L* is placed between a screen and a very small bulb as shown at right.

   a. On the diagram, sketch what you would see on the screen when the bulb is turned on.

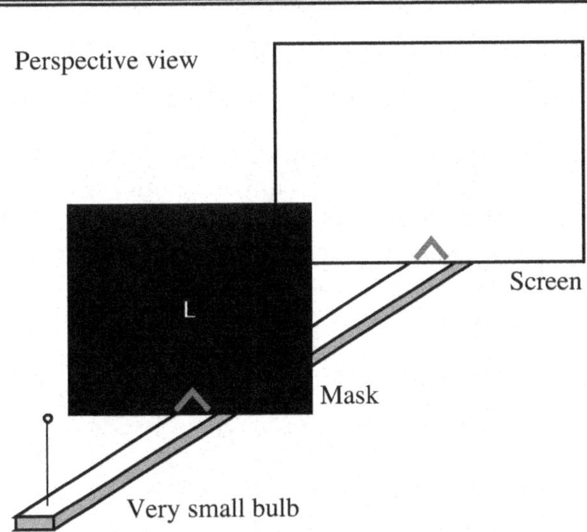

   b. The small bulb is replaced by three long-filament light bulbs that are arranged in the shape of the letter *F* as shown at right.

      On the diagram, sketch what you would see on the screen when the bulbs are turned on. The scale of your sketch should be consistent with your answer to part a. *Explain how you determined your answer*.

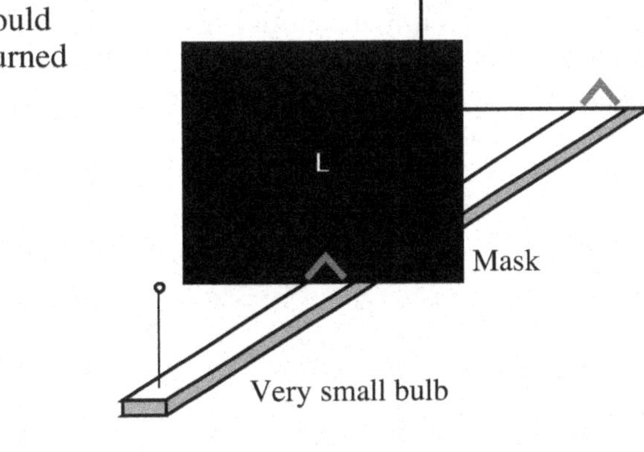

   c. The three long-filament bulbs are replaced by a small bulb and a large triangle-shaped bulb as shown at right.

      On the diagram, sketch what you would see on the screen when the bulbs are turned on. The scale of your sketch should be consistent with your answer to part a. *Explain how you determined your answer*.

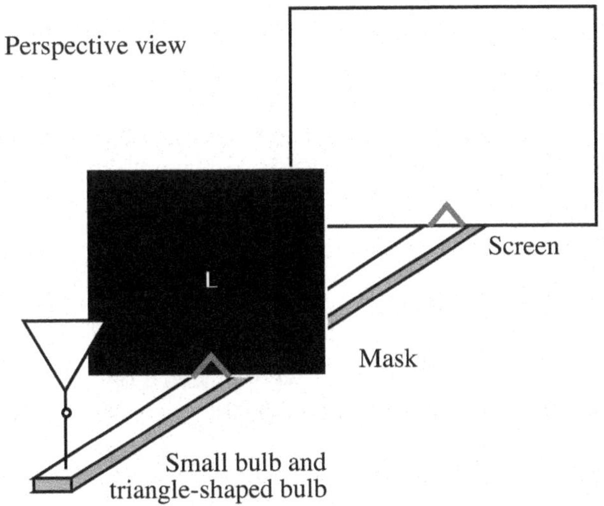

2. A piece of cardboard has been cut into the shape of a triangle. The dimensions of the triangle are shown at right.

20 cm

20 cm

   a. Predict the size and shape of the shadow that will be formed on the screen when a lit bulb, the cardboard triangle, and a screen are arranged along a line as shown. Assume the bulb is small enough that it can be regarded as a point source of light. Explain your reasoning.

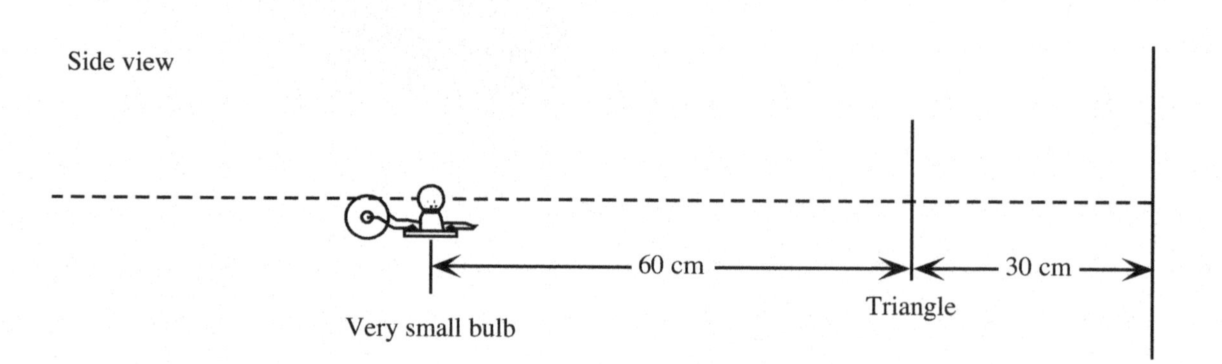

   Sketch your prediction in the space at right.

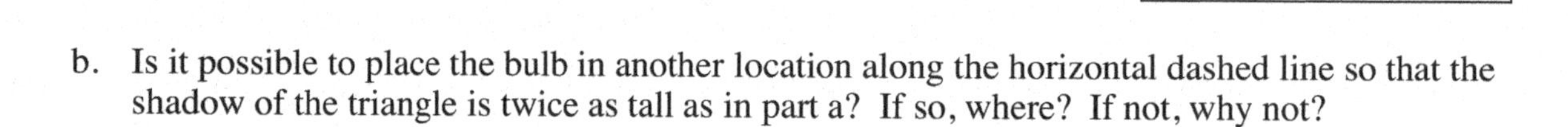

   b. Is it possible to place the bulb in another location along the horizontal dashed line so that the shadow of the triangle is twice as tall as in part a? If so, where? If not, why not?

   c. Is it possible to place the bulb in another location along the dashed line so that the shadow is half as tall as in part a? If so, where? If not, why not?

   d. Suppose that the bulb were placed along the dashed line very far away from the triangle and the screen. What would be the approximate shape and size of the shadow? Explain.

3. A *pinhole camera* can be made as follows. A small circular hole (*e.g.*, about 2 mm in diameter) is made in the center of one side of a box. A larger eyehole is made near one end of the box on the same side as the pinhole. (See perspective view below.) While a person is looking into the box, light can enter only through the pinhole, as shown in the top view diagram.

Perspective view of camera

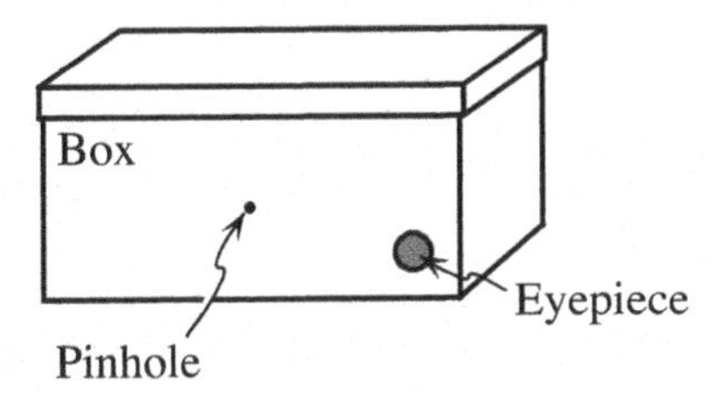

Top view of camera and observer

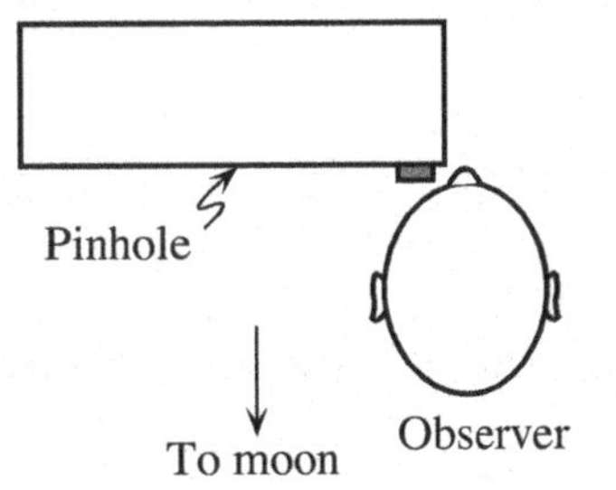

a. Imagine that you viewed a crescent moon with the pinhole camera. The figure at right shows the lit portion of the moon as it would appear in the sky.

i. In the space at right, sketch what you would see inside the camera. (*Hint:* You would have to turn your back to the moon in order to view it with the pinhole camera. A diagram showing the moon, the pinhole, you, and the screen may help you predict the correct orientation of the image that you would see.) Explain your reasoning.

Image in pinhole camera with 2-mm-diameter pinhole

ii. Suppose that the pinhole were larger or smaller than the one above. In the spaces provided, sketch the image if the pinhole were 3 mm or 1 mm in diameter. Explain.

Image in pinhole camera with 3-mm-diameter pinhole

Image in pinhole camera with 1-mm-diameter pinhole

iii. What advantage(s) would there be to using (1) a larger pinhole? (2) a smaller pinhole? Explain.

b. A student is looking at the building shown at right. The student then turns around and views the building through a pinhole camera. In the space below, sketch what the student would see in the camera. Explain your reasoning.

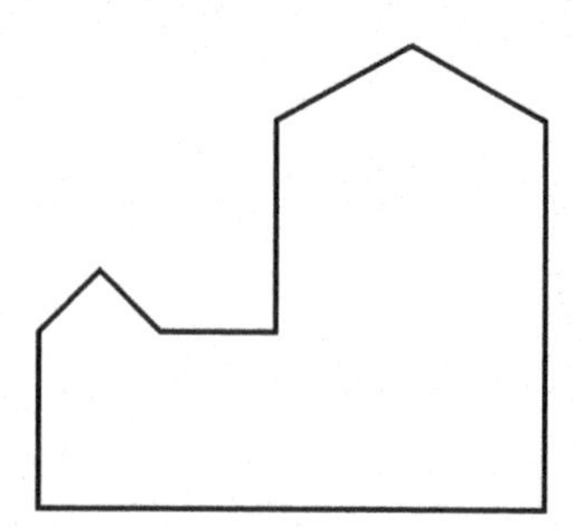

4. A physics student strolling through a park on a sunny day walks under some trees and sees bright circular spots on the ground.

   a. The student, who has been studying geometrical optics in physics class, comes up with the following explanation to account for the shape of the bright spots.

      "The sunlight is passing through holes in the canopy of leaves above my head and forming the bright spots. The bright spots on the ground are circular in shape, so the spaces between the leaves must also be circular."

      Do you agree or disagree with this statement? Explain your reasoning.

   b. Suppose that this student were walking through the same park during an eclipse of the sun, so that the sun was crescent shaped, rather than round.

      What would the student see on the ground in this case? Would it be different than what the student observed on the sunny day? If so, how? If not, why not?

NOTE: You will need a *straightedge* and a *protractor* for this homework.

1. The top view diagrams at right were drawn by a student who is studying image formation by a plane mirror. Each diagram shows the location of an object and two lines of sight to the image of that object in the mirror.

   For each diagram, determine whether or not the situation illustrated is possible. If a situation is possible, draw the location and orientation of the mirror.

   Explain how you reached your conclusions.

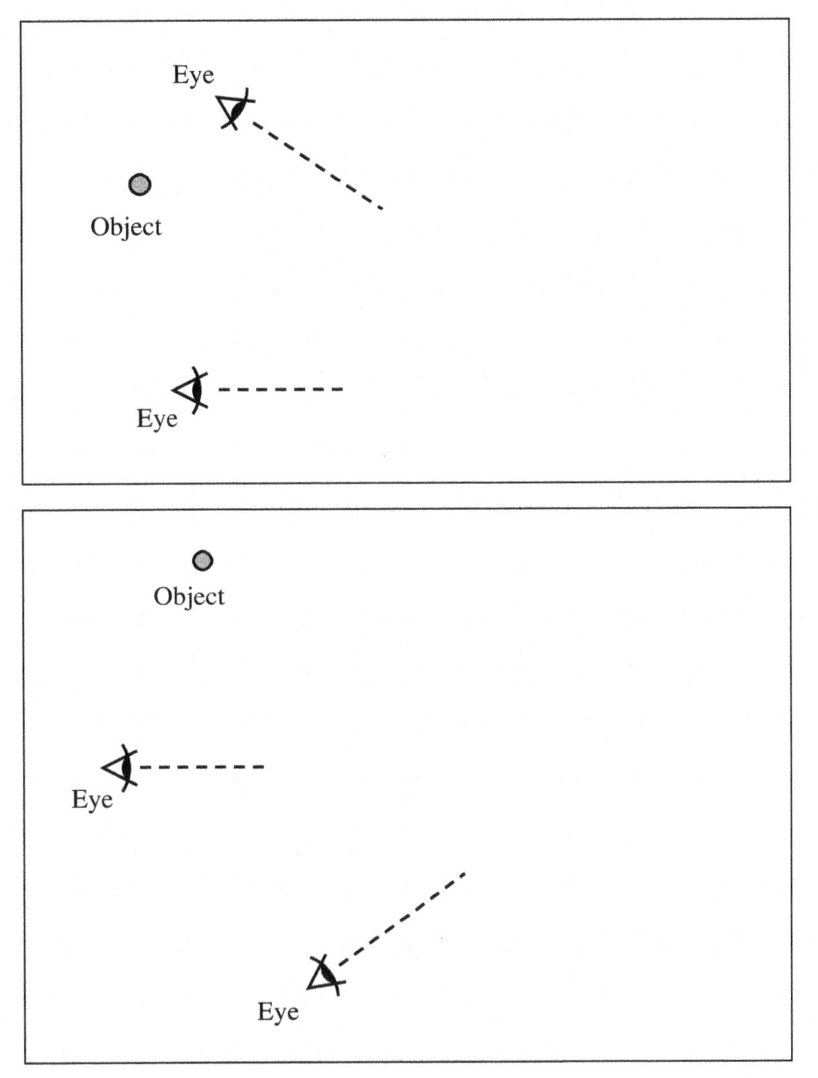

2. Two small objects and a mirror are arranged as shown below.

   a. Draw a ray diagram to determine the location of the image of each of the two objects.

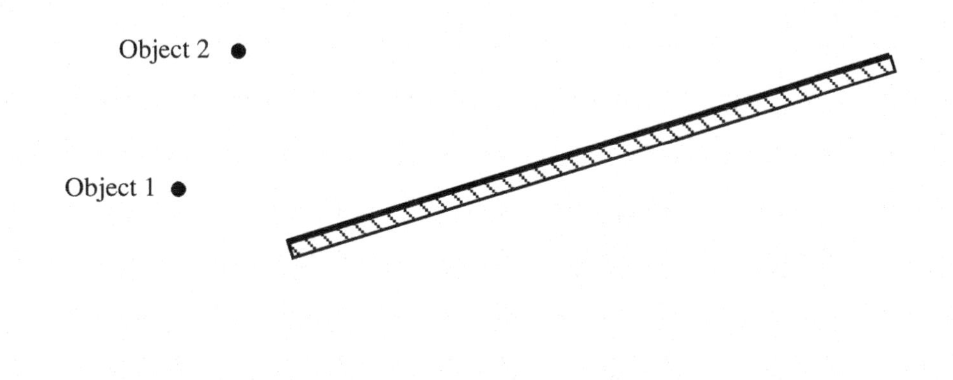

b. Describe how you could use a ray diagram to determine the location of the image of an extended object, such as a pencil.

3. A pencil is placed in front of a plane mirror as shown in the top view diagram below.

a. Use ray tracing to determine the location of the image of the pencil. Use a protractor and a straightedge to make an accurate drawing.

Clearly indicate the entire image on your diagram.

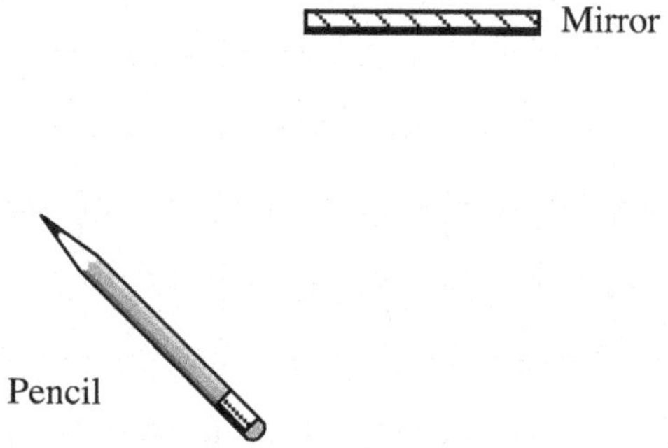

Mirror

Pencil

b. Draw a ray diagram below to determine the region in which an observer must be located to see (i) the image of the tip of the pencil, (ii) the image of the eraser, and (iii) the image of the entire pencil. Clearly label each region on your diagram. Explain your reasoning.

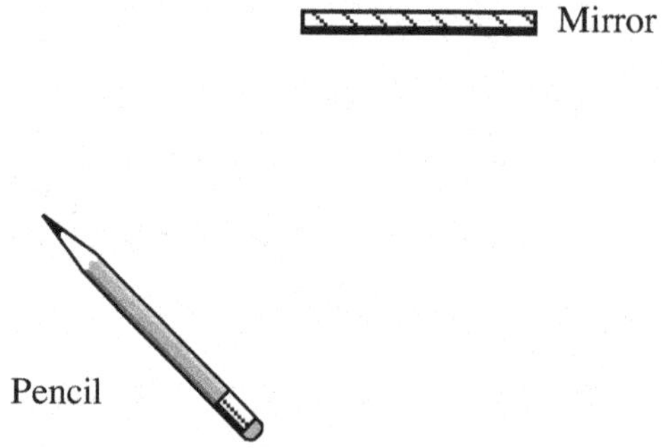

Mirror

Pencil

Note: You will need a *straightedge* and a *protractor* for both problems on this homework.

1. A pin is placed in front of a semi-cylindrical mirror as shown in the top view diagram below.

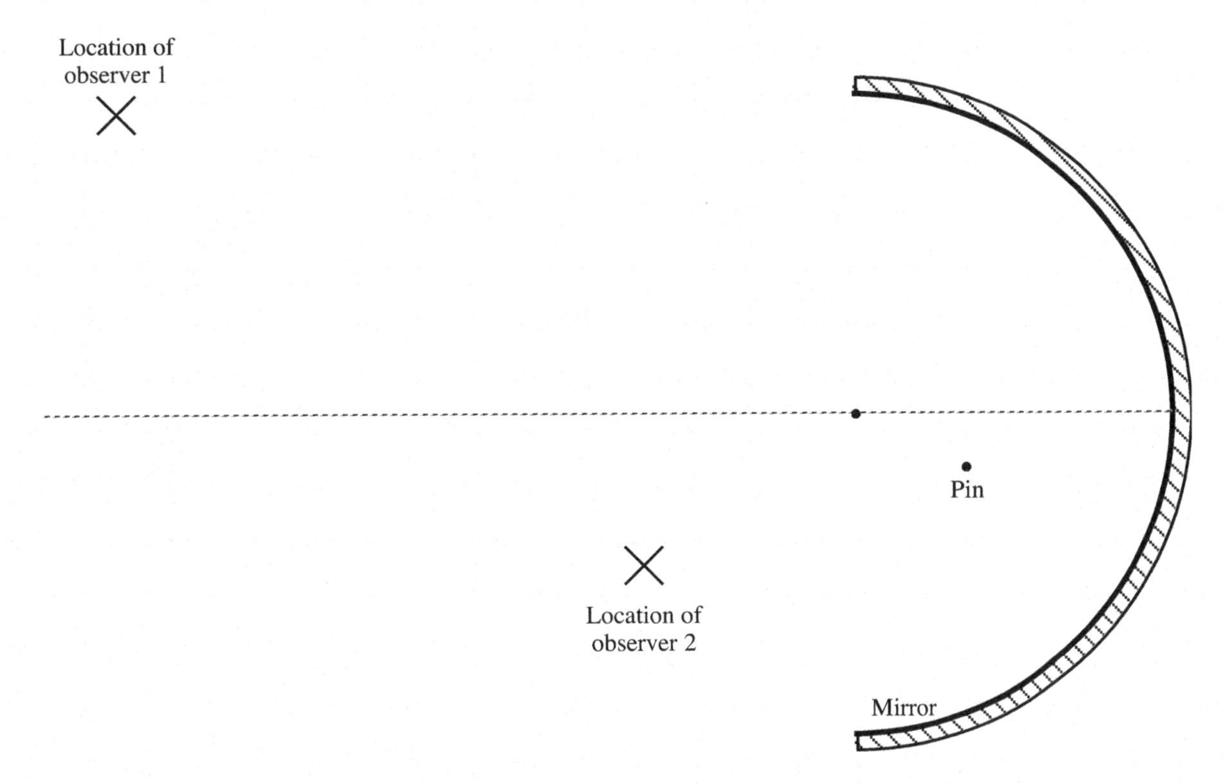

a. Use the law of reflection to draw a ray diagram that shows the location of the image seen by observer 1. Use a protractor and a straightedge to make your diagram as accurate as you can. (*Note:* The center of curvature and the axis of the mirror are marked.)

   Clearly label the location of the image on your ray diagram.

b. Is this image *real* or *virtual?* Explain your reasoning.

c. Will observers 1 and 2 agree on the location of the image of the pin? Support your answer with a ray diagram and explain how you used the diagram to determine your answer.

2. A very small, very bright bulb is placed far from a semi-cylindrical mirror. The bulb is located on the axis of the mirror. Some light rays from the distant bulb are shown in the diagram below.

To small bulb,
far from mirror

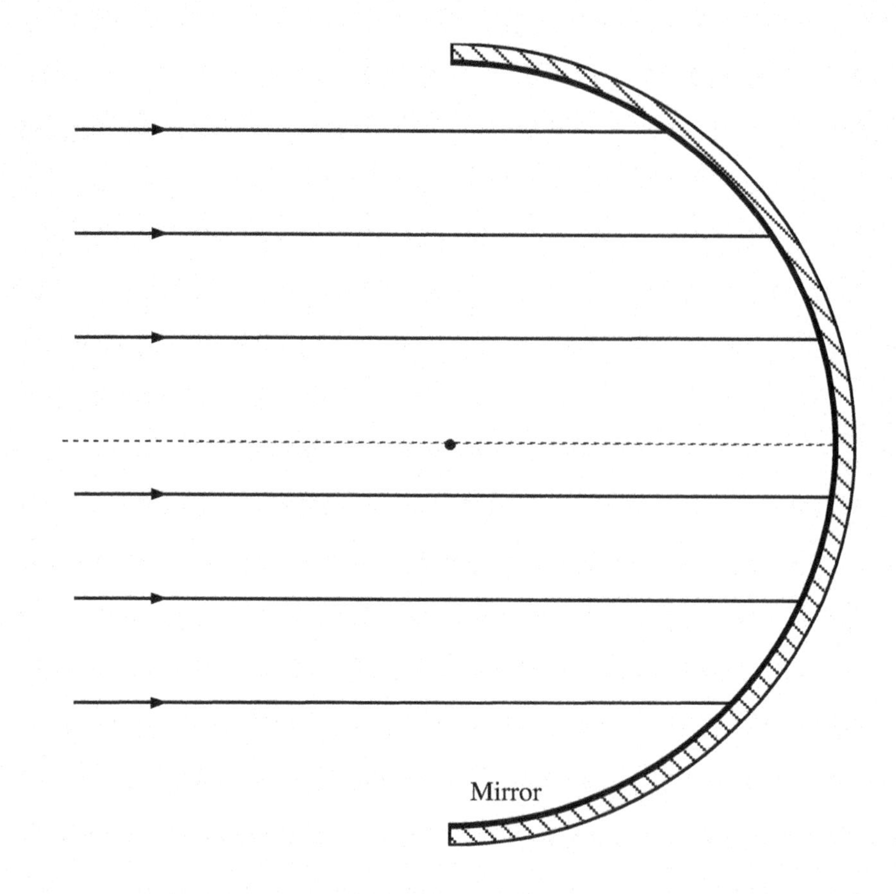

a. Use the law of reflection to continue the rays shown in the diagram. Use a protractor and a straightedge to make your diagram as accurate as you can. (*Note:* The center of curvature and the axis of the mirror are marked.)

b. Is there a well-defined focal point for the (entire) mirror?

If so: Identify and label the focal point on the diagram. Explain how you used your ray diagram to determine your answer.

If not: Identify and label the approximate portion of the mirror for which a focal point is well defined. Identify and label the focal point for this portion of the mirror. Explain how you used your ray diagram to determine your answers.

INTERPRETATION OF RAY DIAGRAMS

Name ____________________

1. The following are top view diagrams of solid cylinders and cubes. Assume that light travels more slowly through the objects than through the surrounding medium.

   Each diagram shows a path for light that is *not* qualitatively correct; there is at least one flaw, perhaps more, in each diagram. Identify *all* flaws. Explain your reasoning.

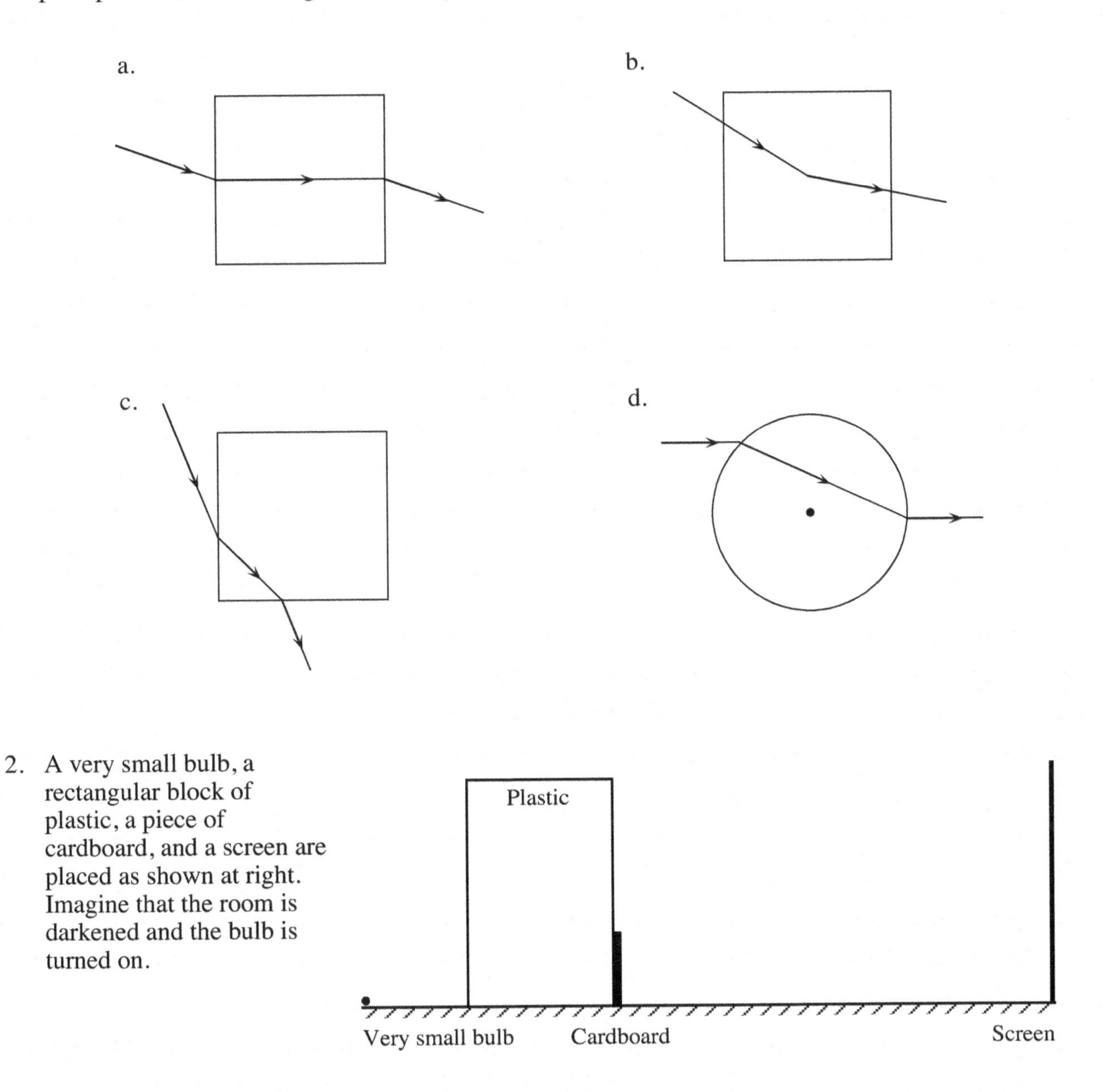

2. A very small bulb, a rectangular block of plastic, a piece of cardboard, and a screen are placed as shown at right. Imagine that the room is darkened and the bulb is turned on.

   If the plastic were removed, would the height of the shadow on the screen *increase, decrease,* or *stay the same?* Explain your reasoning, and support your answer with a clear ray diagram.

3. A nail is placed in a tank of water as shown in the top view diagram below. (Only a portion of the tank is shown.) Assume that light passes directly from water to air.

   a. Use a protractor and straightedge to draw each of the four refracted rays on the diagram *accurately*. (The index of refraction for water is 1.33.) Record the angles of incidence and refraction in the table below.

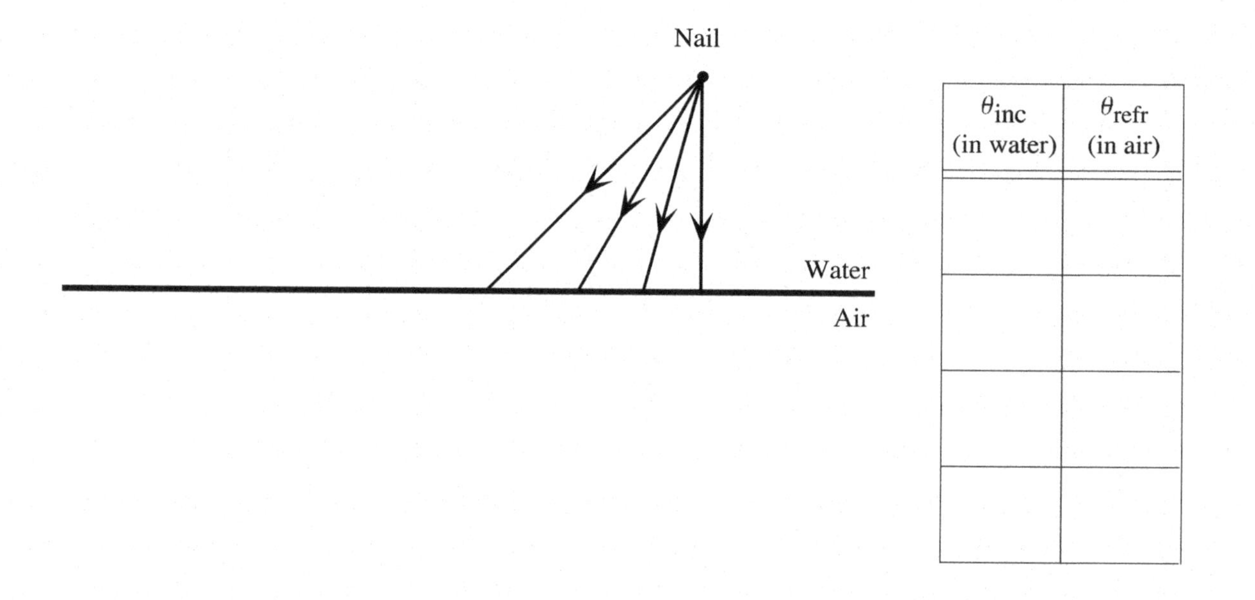

θ_{inc} (in water)	θ_{refr} (in air)

   b. If several different observers were standing in front of the tank, would they agree upon the location of the image of the nail (as viewed through the water)? Explain how you can tell from your ray diagram.

      - If different observers would *agree* upon the image location, indicate this image location on the diagram.
      - If different observers would *not agree* upon the image location, indicate on the diagram the approximate image location for each of *three* different observer locations. Clearly indicate which image location corresponds to which observer location.

   c. Is the image(s) of the nail *real* or *virtual?* Explain your reasoning.

CONVEX LENSES

Name ______________

1. A small bulb is placed in front of a convex lens.

    a. Suppose that the bulb is placed as shown. Using all three principal rays, draw an accurate ray diagram to determine the location of the image. Label the image location.

    b. Repeat part a for the case shown at right, in which the bulb is farther from the lens.

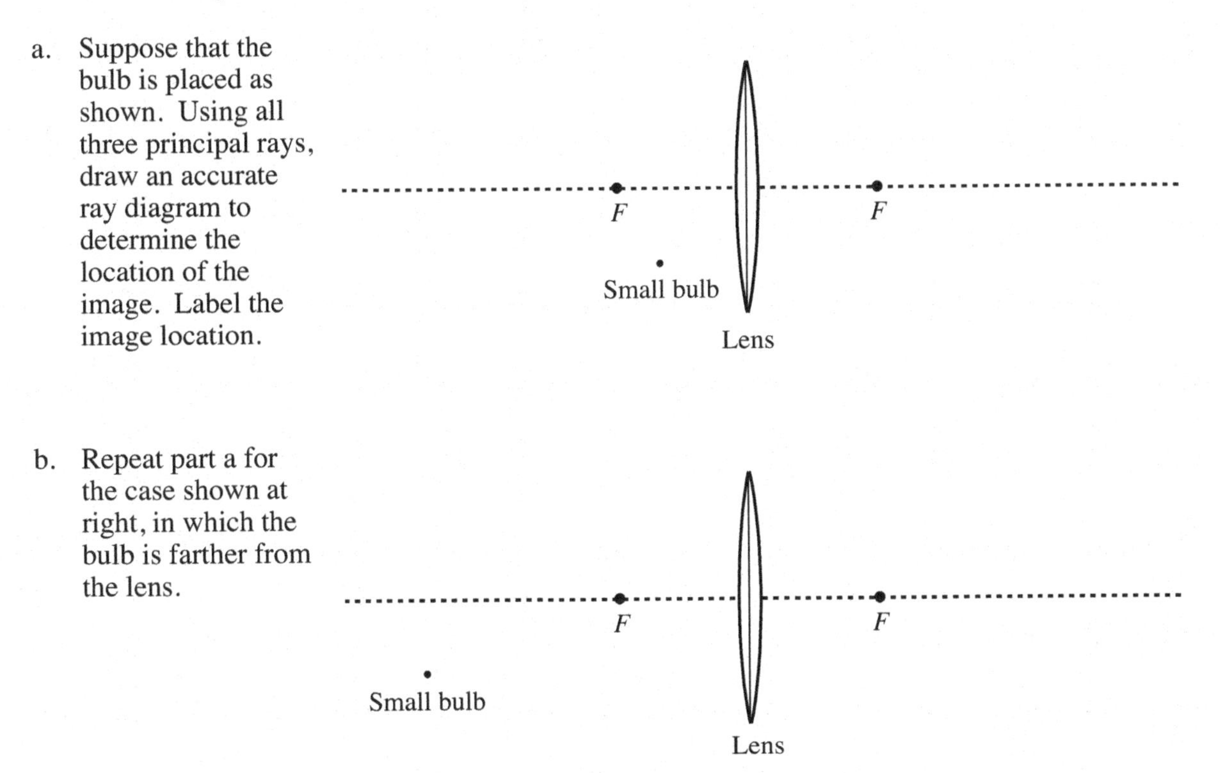

    c. Suppose that in each case above you were to place a small paper screen at the image location. What you would see on the screen in each case? Imagine that the room is dark except for the small bulb.

    d. The light bulb is placed at one focal point of the lens as shown at right.

    Draw at least five rays from the bulb that pass through the lens.

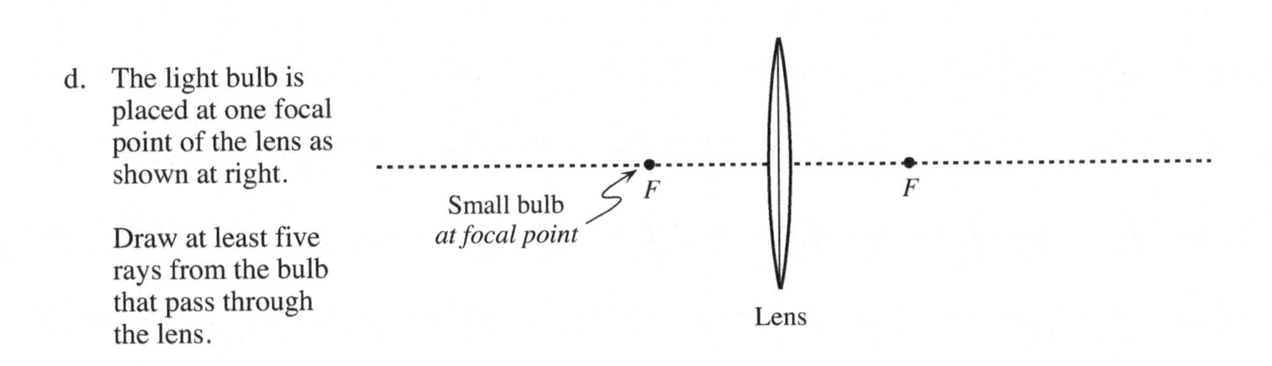

    Where is the image located in this case? Explain. (*Hint:* How are the rays that have passed through the lens oriented? From where do these rays *appear* to have come?)

2. Two thin convex lenses (1 and 2) and a small object are arranged as shown.

   a. Use the three principal rays to determine the location of the image of the object produced by lens 1.

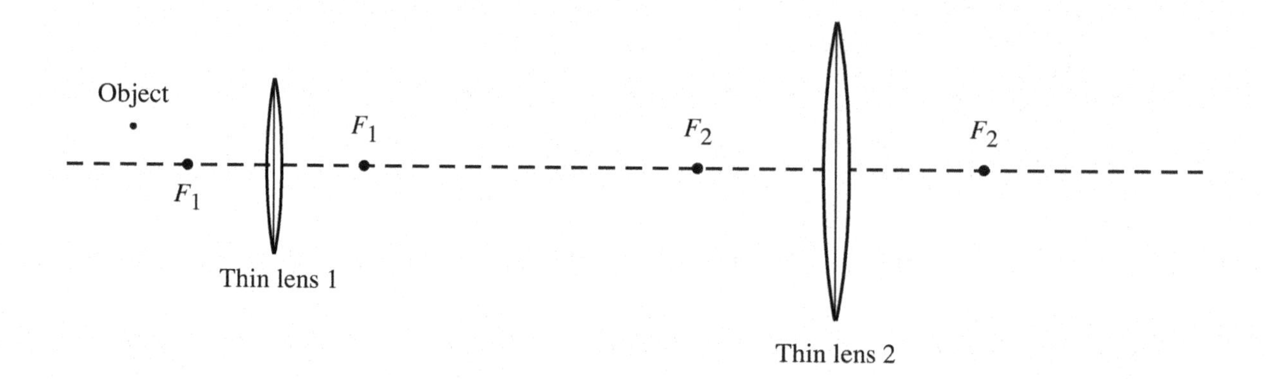

   b. Treat the image produced by lens 1 as an object for lens 2. Use the three principal rays to determine the location of the image of this object produced by lens 2.

   Is this image produced by the pair of lenses *real* or *virtual?* Explain your reasoning.

   c. Repeat parts a and b for the case in which lens 2 is replaced with a different lens (lens 3), as shown below.

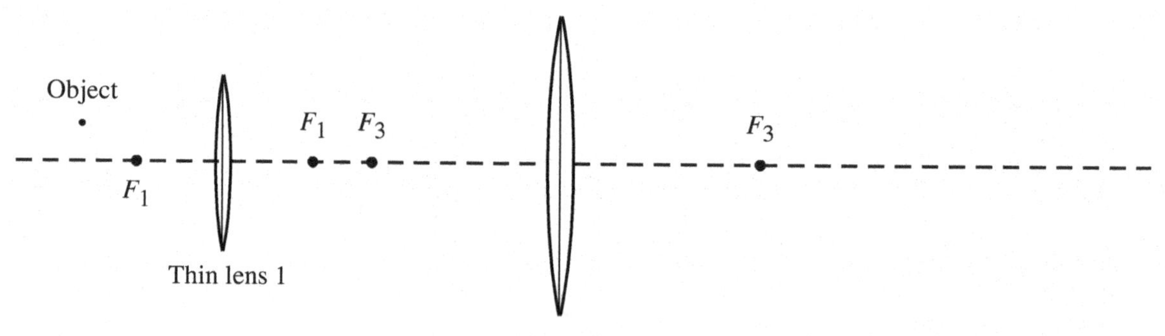

   Is the image produced by the pair of lenses *real* or *virtual?* Explain your reasoning.

1. Reproduced below is a side view diagram of the situation described in section II of the tutorial.

   Determine an expression for the lateral magnification, $m_l = h'/h$, in terms of the object distance, x_o, and the image distance, x_i. (*Hint:* Draw the principal rays for the tip of the pencil and look for similar triangles. Clearly indicate the similar triangles that you use to determine your answer.)

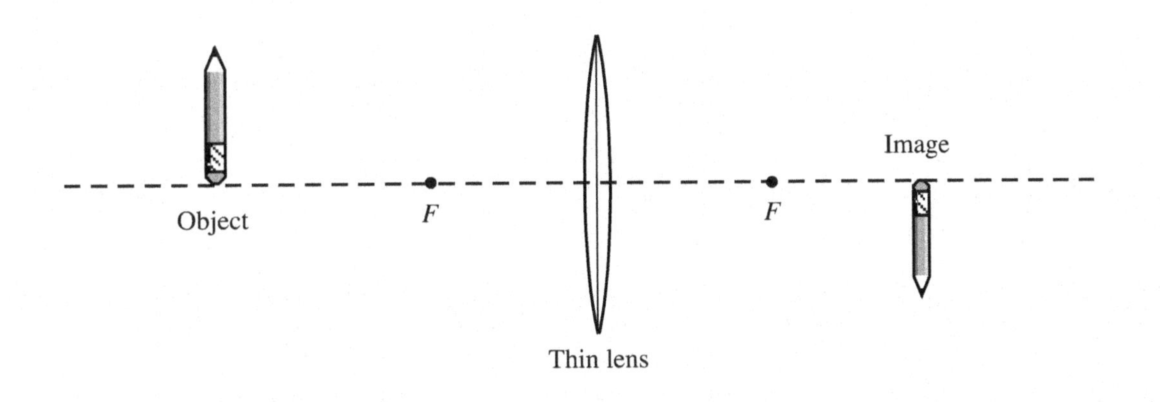

2. In section III of the tutorial *Magnification,* you used a convex lens as a magnifying glass.

   Is the expression that you derived in problem 1 above for the lateral magnification, m, also valid in this case? If not, what expression holds in this case? Draw a diagram below to support your answer.

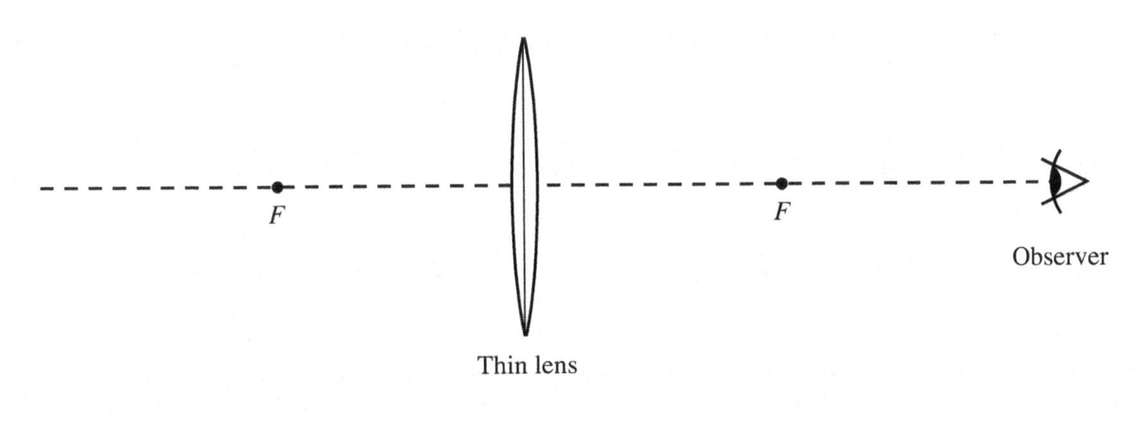

3. Two thin convex lenses and an object are arranged as shown below. Two rays from the tip of the object are drawn in order to determine the location of the image produced by lens 1. Lens 2 is placed so that one of its focal points coincides with the location of the image produced by lens 1.

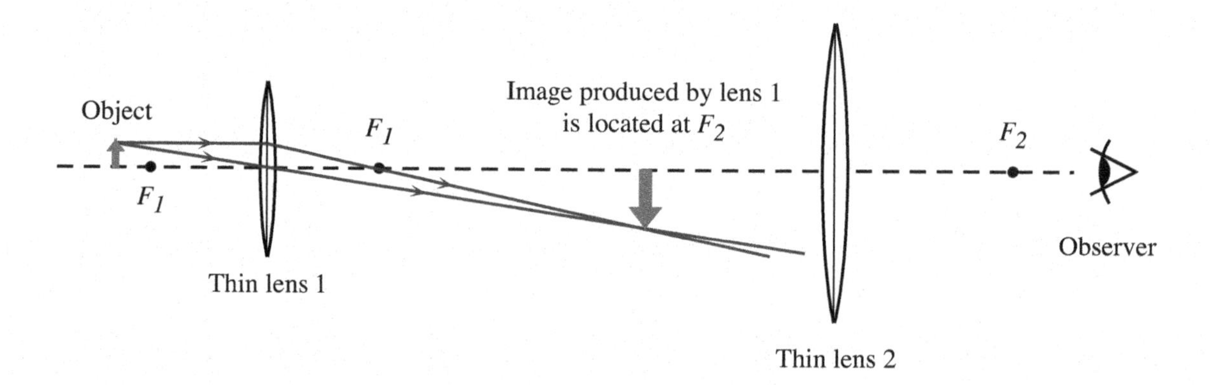

a. Treating the image produced by lens 1 as an object for lens 2, draw *two* principal rays from the tip of this image that pass through lens 2. (Note that one of the principal rays cannot be drawn in this case.)

Using either geometry or trigonometry, show that these two principal rays are *parallel* on the right side of lens 2. (*Hint:* Look for congruent right triangles in your ray diagram.)

b. Where is the tip of the image seen by the observer located? Explain. (*Hint:* From where do the rays on the right side of lens 2 *appear* to have come?)

c. On the diagram above, clearly indicate:

- the direction in which the observer must look to see the *tip* of the image,
- the direction in which the observer must look to see the *tail* of the image, and
- an angle that represents the angular size of the *entire* image seen by the observer.

d. On the basis of your results in parts a–c, which would *appear* larger: the image seen by the observer (with both lenses present) or the object itself (with the lenses removed)? Explain.

e. The diagram in this problem illustrates a *compound microscope*. Lens 1, called the *objective,* is placed near the object of interest. Lens 2, known as the *eyepiece,* is placed so that one of its focal points coincides with the image produced by the objective.

In order to improve the angular magnification of the microscope shown above, would you replace the eyepiece (lens 2) with another lens that has a *smaller* focal length or a *larger* focal length? Explain your reasoning.

TWO-SOURCE INTERFERENCE

Name ____________________

1. The top view diagram at right illustrates two point sources, S_1 and S_2.

   a. On the diagram, indicate points for which the value of ΔD is (i) largest and (ii) smallest. (ΔD is the difference in distances to the sources.)

   b. What are the largest and smallest values of ΔD for this situation? Explain your reasoning.

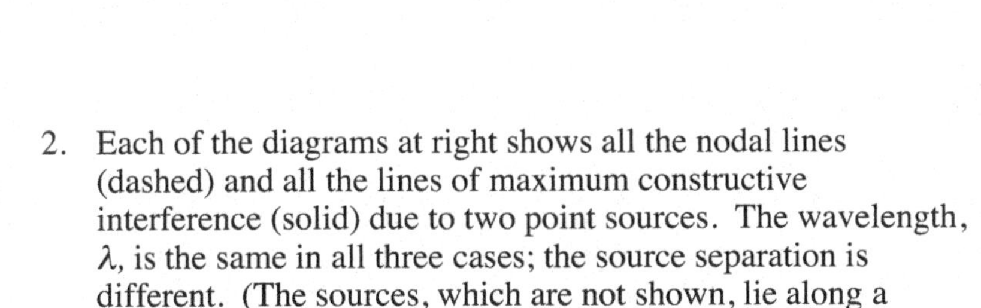

2. Each of the diagrams at right shows all the nodal lines (dashed) and all the lines of maximum constructive interference (solid) due to two point sources. The wavelength, λ, is the same in all three cases; the source separation is different. (The sources, which are not shown, lie along a horizontal line.)

   a. Label each nodal line and line of maximum constructive interference in the shaded region with the appropriate value of ΔD (in terms of λ).

   b. For each case, determine the source separation (in terms of λ). For any case(s) for which it is not possible to determine the source separation *exactly,* determine the source separation as closely as you can, *i.e.,* give the smallest range into which the source separation must fall. (*Hint:* You may find it helpful to first rank the cases by source separation.)

   Explain how you determined the source separations.

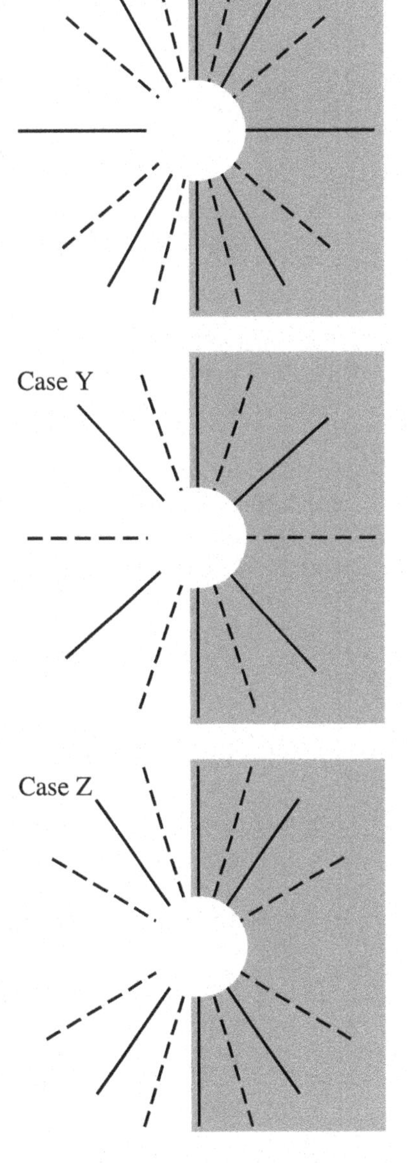

3. The diagram at right shows an arbitrary point, point A, that lies near two point sources of waves. In this problem, we consider how the phase difference at point A changes as point A is moved outward along the dark line, away from the sources.

   Point Z on the diagram below is the same distance from point A as is the source S_2.

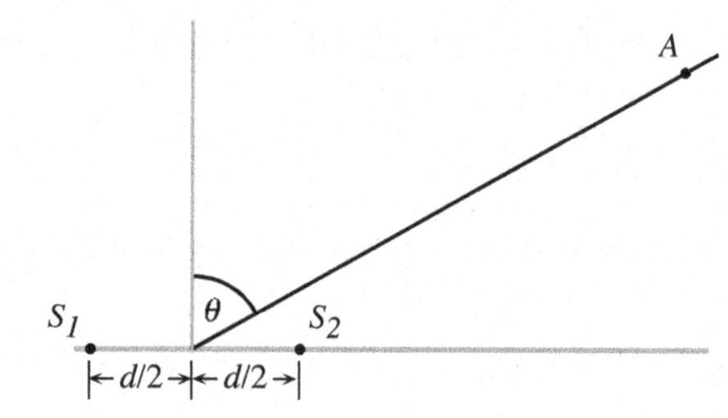

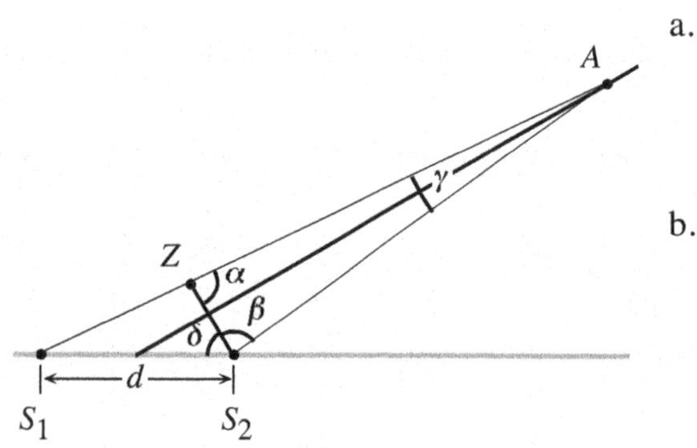

   a. How do the angles α and β compare? Explain.

   b. Suppose that point A is moved away from the sources along the dark line. In the limit that point A is very far from the sources, what values do the angles α, β, and γ approach?

   c. On the diagram above, indicate the line segment that represents how much farther point A is from S_1 than it is from S_2. Label this distance ΔD.

   d. The enlarged diagram at right illustrates the limit in which point A is moved very far from the sources.

      In this limit, find an expression for ΔD in terms of the angle θ and the source separation d. (*Hint:* How do the angles δ and θ compare?)

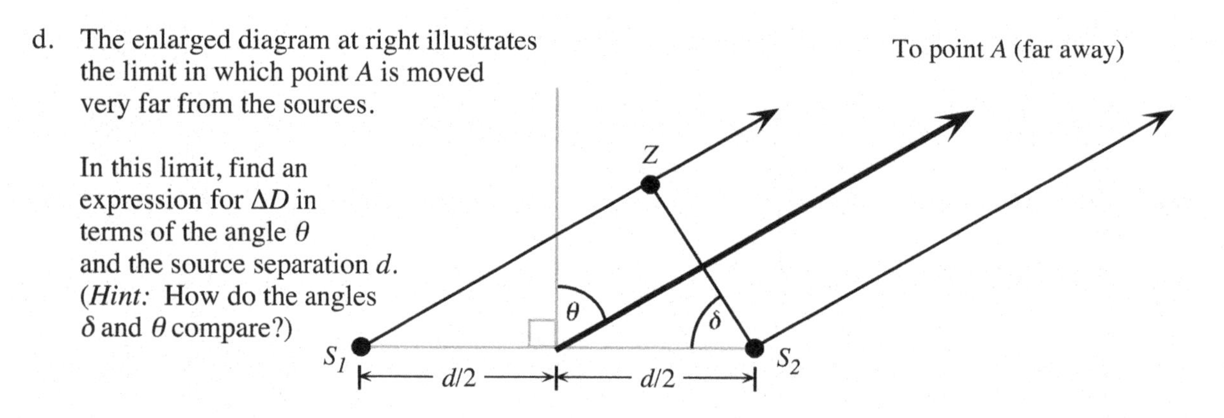

   e. For what values of ΔD (in terms of λ) will there be:

      - maximum constructive interference?

      - complete destructive interference (*i.e.*, a node)?

f. Use your answers from parts d and e to write equations that can be used to determine the angle(s) for which there will be:

- lines of maximum constructive interference

- nodal lines

g. Determine the angles for which there will be nodal lines and lines of maximum constructive interference for the case of two sources in phase, a distance 1.5λ apart.

   i. Use your results to draw *accurate* nodal lines and lines of maximum constructive interference on the diagram at right.

2 sources in phase
$d = 1.5\lambda$ between sources (not shown)

   ii. Label each line from part i with the corresponding value of ΔD, θ, and $\Delta\varphi$, where $\Delta\varphi$ is the phase difference between the waves.

   iii. If the distance between the sources were increased, would the angle θ to the first nodal line *increase, decrease,* or *stay the same?* Explain your reasoning.

4. Consider the following *incorrect* statement referring to problem 3:

   "As Point A moves farther and farther away from the sources, the distances to the sources become more nearly equal, so the difference in distances is negligible. Thus the waves are more nearly in phase as Point A moves farther and farther away from the sources."

   What is the flaw in this argument? Explain your reasoning.

1. A distant point source of red light, a mask with two identical, very narrow slits, and a screen are arranged as shown in the top view diagram below right.

   The photograph at right shows the pattern that appears on a distant screen. Point *P*, the center of the pattern, and point *Q* are maxima. Point *R* marks a minimum to the right of point *Q*.

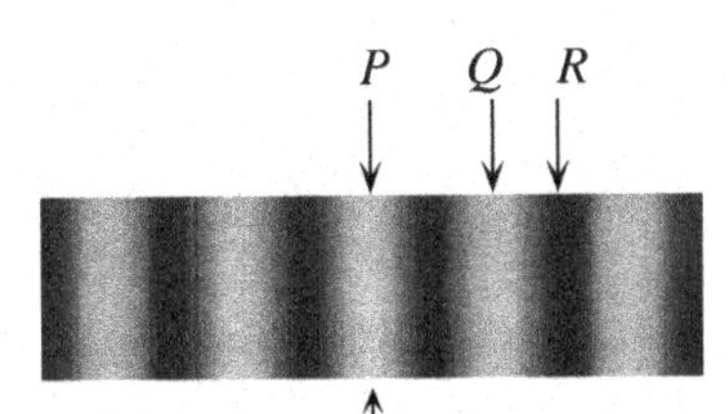

   a. In the space above the photograph at right, clearly label each of the lettered points according to ΔD, the difference in distances from the slits to that point. Express each value of ΔD in terms of λ.

   b. The screen is 2.2 m from the slits, and the distance from point *P* to point *R* is 1.6 mm.

      Determine the distance between the slits in terms of λ. Show your work and describe any approximations that you make in answering this question.

Top view (*not* to scale)
P
Screen
Mask with 2 slits
To distant point source

   c. Suppose that the width of the right slit were decreased (without changing the distance between the centers of the slits).

      i. Would the intensity at each of the following points *increase, decrease,* or *stay the same?* In each case, explain your reasoning.

         - point *Q*

         - point *R*

      ii. The graph of intensity *(I) versus* angle *(θ)* shown at right corresponds to the above double-slit experiment. (The angle θ is measured relative to the normal to the screen.)

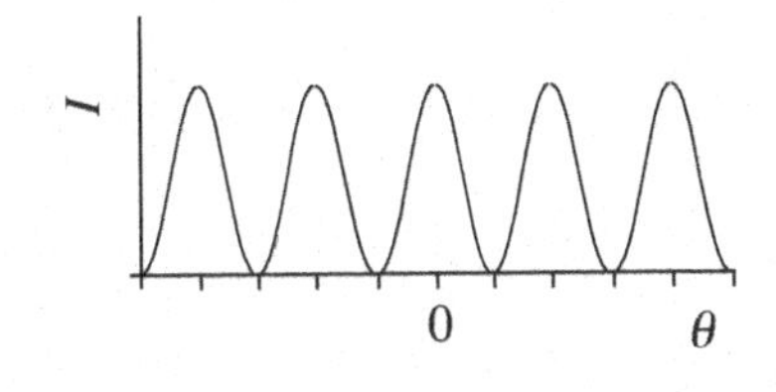

          In the space at right, show how the intensity graph would be different if the right slit were made narrower.

2. The graph of intensity *versus* angle at right corresponds to a double-slit experiment similar to the one described in problem 1.

In each part below, suppose that a *single* change were made to the original apparatus. In the spaces provided, show how the graph of intensity *versus* angle would be different from the original graph (shown dashed). In each case, *explain your reasoning*.

- the distance between the centers of the slits is increased (without changing the width of the slits)

- the wavelength of the incident light is increased

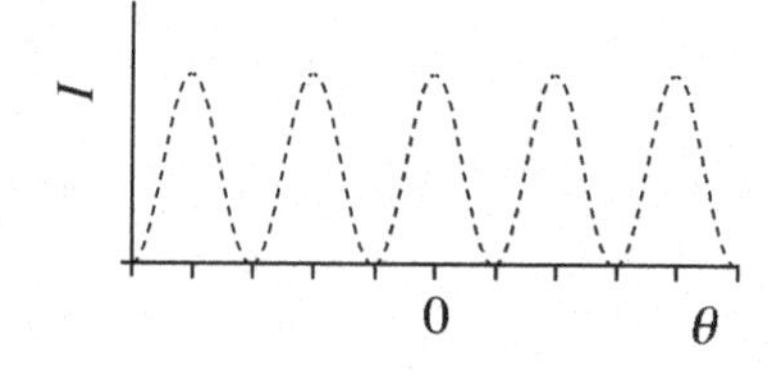

- the distance from the mask to the screen is decreased

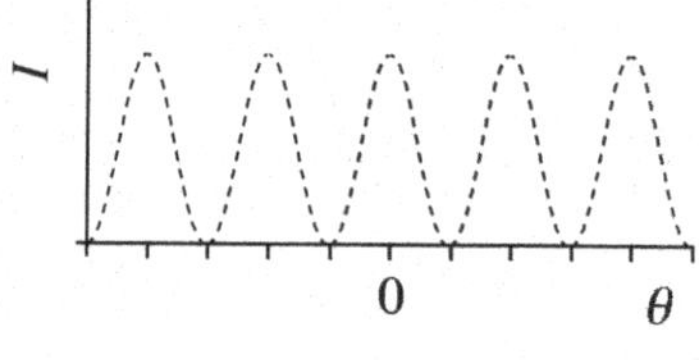

- the width of *both slits* is decreased (without changing the distance between the centers of the slits)

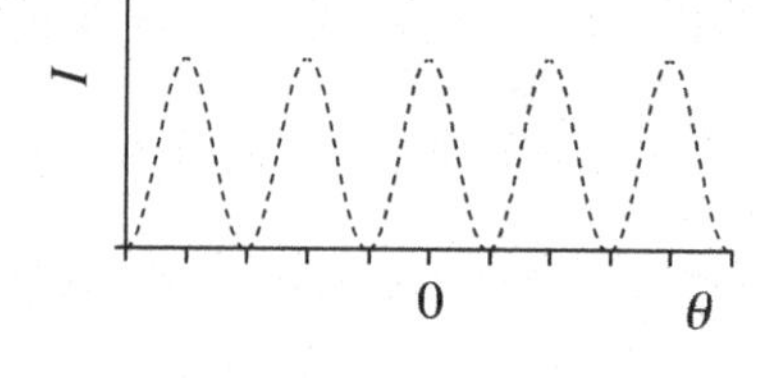

- the width of *one* of the slits is decreased (without changing the distance between the centers of the slits) [Sketch the approximate shape of the intensity *versus* angle graph.]

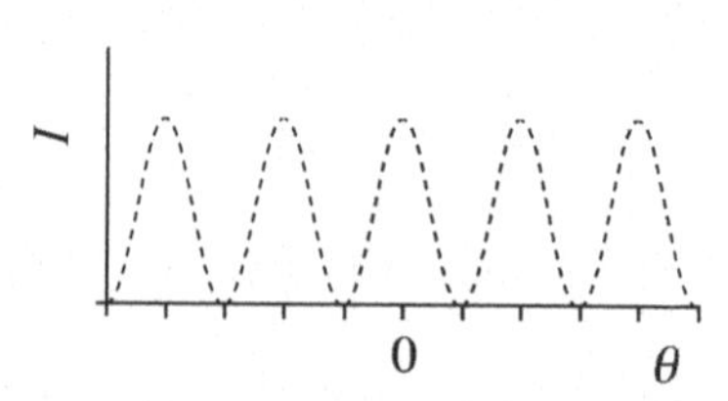

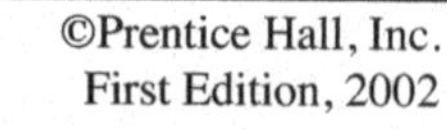

MULTIPLE-SLIT INTERFERENCE

Name ______________________

1. The photograph at right illustrates the pattern that appears on a distant screen when coherent red light is incident on a mask with two identical, very narrow slits. The two slits are separated by a distance d.

   Pattern on screen with two slits

   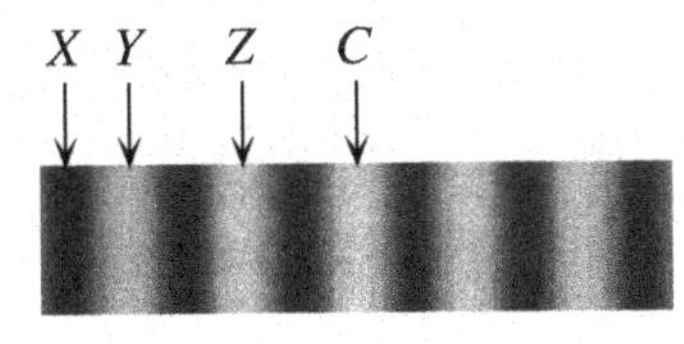

   Point C marks the center of the screen. Point X marks a point of zero intensity, and points Y and Z mark maxima.

   a. In the space at right below the photograph, clearly label each of the lettered points according to ΔD, the difference in distances from the slits to that point. Express each value of ΔD in terms of λ.

   b. Suppose that the two-slit mask were replaced by a mask with *five* slits of the same width as the original slits, with adjacent slits separated by the same distance as before.

      i. Would point X still be a point of zero intensity with the five-slit mask? Explain your reasoning.

      ii. Would point C, the center of the screen, still be a point of maximum constructive interference with the five-slit mask? Explain your reasoning.

      With the five-slit mask, how many minima would there be between each pair of principal maxima?

      iii. Would point Z still be a point of maximum constructive interference with the five-slit mask? Explain your reasoning.

      iv. If the distance between the centers of adjacent slits in the five-slit mask is 125λ, determine the angle (measured relative to the normal to the screen) to the minimum closest to the center of the screen. Show your work.

2. Consider the original double-slit pattern from problem 1, shown at right. Suppose that a third slit of the same width were added halfway between the original two slits as shown in the figure below the pattern. (Note that this results in the distance between adjacent slits becoming *half* of the original value.)

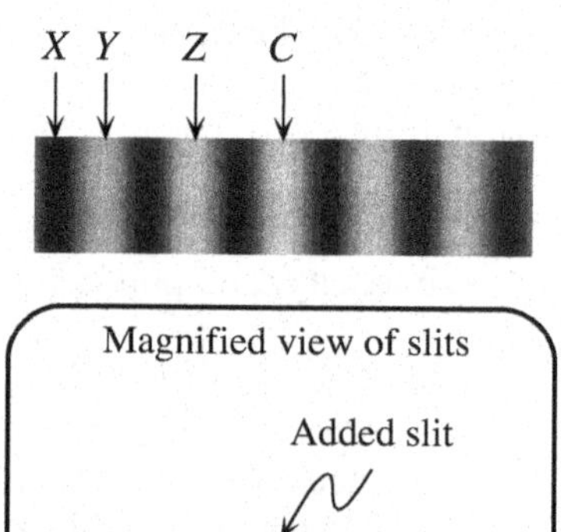

   a. Would point Z be a *principal maximum*, a *minimum*, or *neither?* Explain your reasoning.

   b. Would point Y be a *principal maximum*, a *minimum*, or *neither?* Explain your reasoning.

   c. Would point X be a *principal maximum*, a *minimum*, or *neither?* (*Hint:* Use superposition.) Explain your reasoning.

3. Monochromatic light from a distant point source passes through a mask containing an unknown number of slits. The graph of intensity *versus* angle is shown at right.

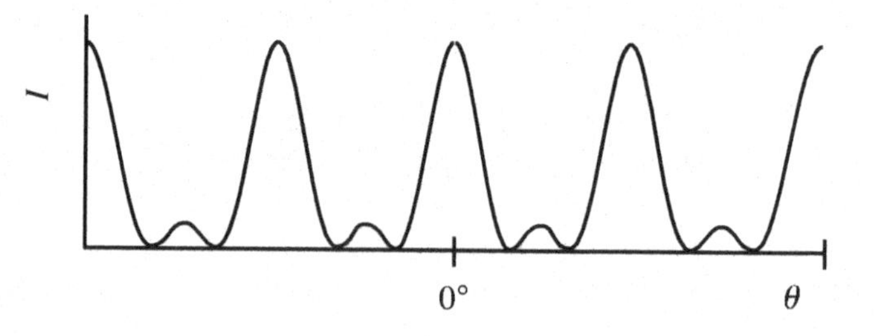

   a. How many slits were there in the mask? Explain how you can tell from the graph. (*Hint:* Base your answer on your results from section III of the tutorial "Multiple-slit interference."

   b. Clearly label each of the minima and principal maxima with the corresponding value of ΔD_{adj} in terms of λ, where ΔD_{adj} is the difference in distances to adjacent slits.

A MODEL FOR SINGLE-SLIT DIFFRACTION

Name ________________

1. Light from a distant point source is incident on a narrow slit. Each of the graphs below shows the intensity on a distant screen as a function of θ. The only difference among the six physical situations is the width of the slit.

   The horizontal scale is the same for all graphs. The vertical scale has been normalized so that the maximum intensity is the same for all cases.

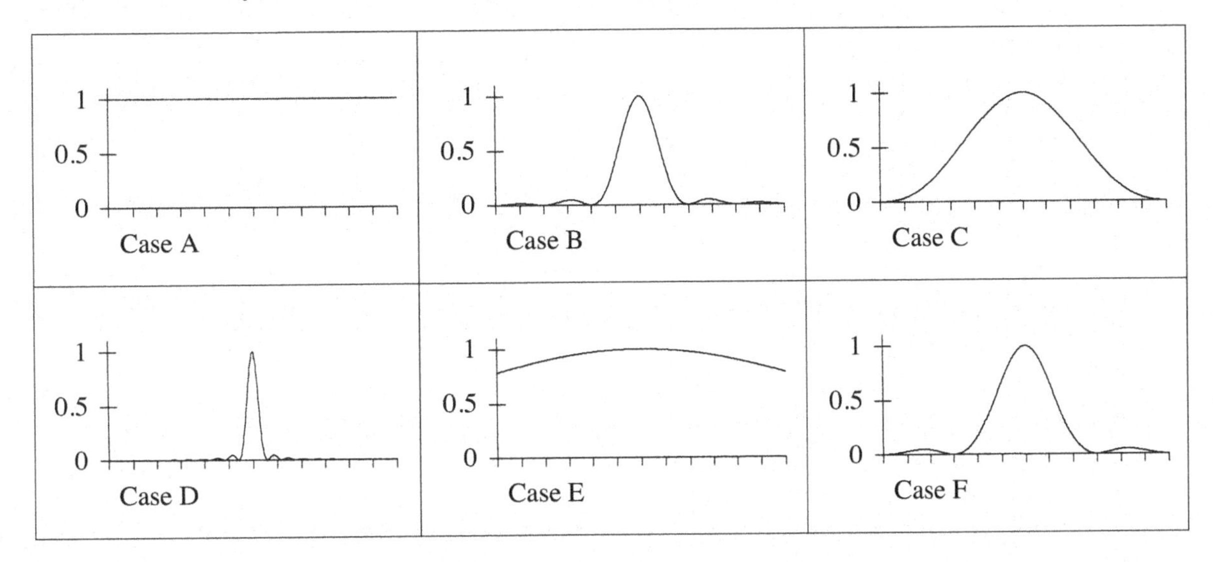

   Rank the cases according to the width of the slit, from largest to smallest. Explain your reasoning.

2. The graph at right shows the intensity on a distant screen as a function of θ for a single-slit experiment.

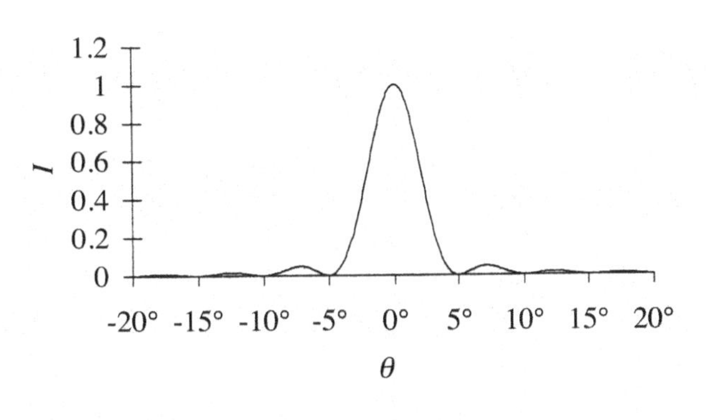

   a. Determine the width of the slit in terms of λ. Show your work.

   b. If $\lambda = 580$ nm, what is the width of the slit in mm? Show your work.

3. There is a systematic way of determining the locations of *all* of the minima in a single-slit diffraction pattern that uses the pairing method developed in sections I and II of the tutorial.

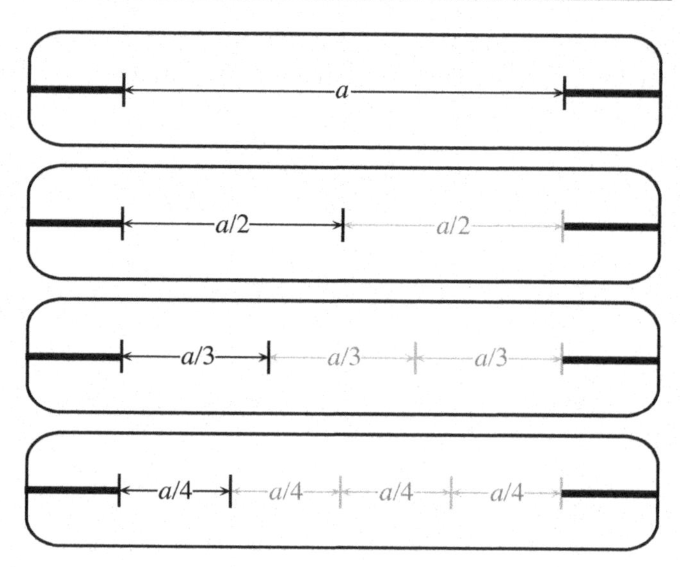

To determine the location of the second minimum, divide the slit in half, then apply the method you have developed in tutorial to each half of the slit. To determine the location of the *third* minimum, divide the slit into *thirds,* then apply the same method to each third of the slit.

In this problem, you will generalize this procedure to find the angle to the mth minimum.

a. If the light that passes through one of the m equal parts yields a minimum at an angle θ for points far from the slit, would the light that passes through each of the other $m - 1$ parts yield a minimum at that same angle? Explain.

b. Use the method described above to write an equation that can be used to determine the angle to:

i. the second minimum

ii. the third minimum

iii. the mth minimum

What values can m take? In particular, can $m = 0$? Explain why or why not.

c. Suppose you were to find *all* of the angles θ at which two very narrow slits separated by $a/2$ would produce minima. How would this set of angles be different from the set of angles given by your equation in part iii above?

1. Monochromatic light from a distant point source is incident on two slits. The resulting graph of *intensity* versus θ is shown below. Point Y is the center of the screen; points X and Z are minima.

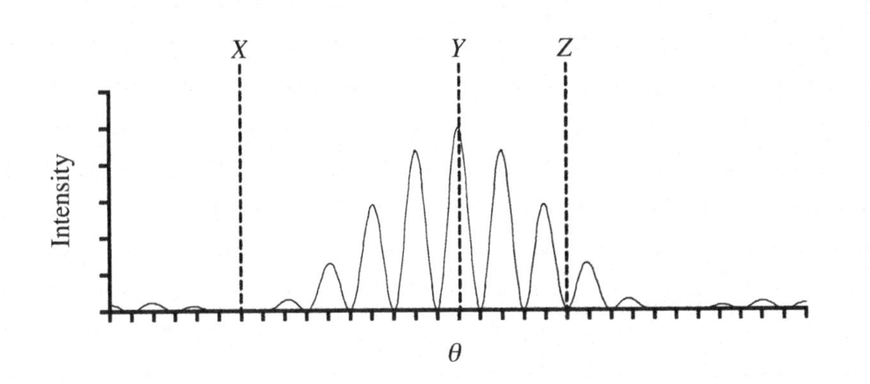

a. If one of the slits in the mask were covered, would the intensity at each of the following points *increase, decrease,* or *stay the same?* Explain your reasoning in each case.

- point X

- point Y

- point Z

b. Determine the ratio of the distance between the slits to the slit width, d/a, for this case. Explain your reasoning. (*Hint:* Can you identify any "missing" interference maxima?)

2. Light from a laser ($\lambda = 633$ nm) is incident on two slits. The resulting graph of *relative intensity* versus θ is shown at right. (The center of the pattern is at $\theta = 0°$.)

Calculate the slit width, a, and the distance between the slits, d. Make clear which features of the graph you use to answer.

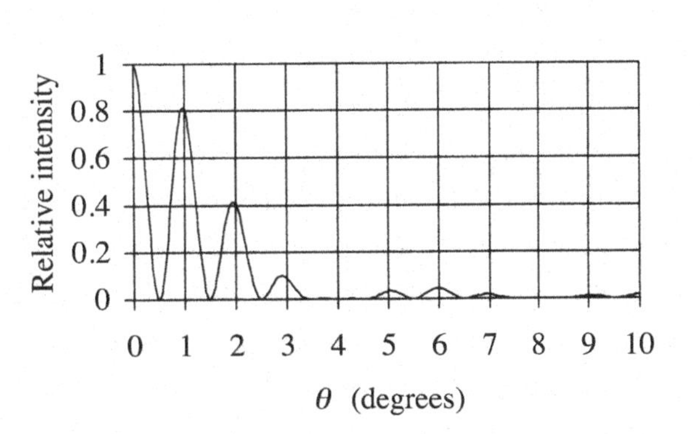

3. Monochromatic light from a distant point source is incident on two slits. The width of each slit is a and the distance between the centers of the slits is d. The resulting graph of *relative intensity* versus θ is shown at right. (The center of the pattern is at $\theta = 0°$.)

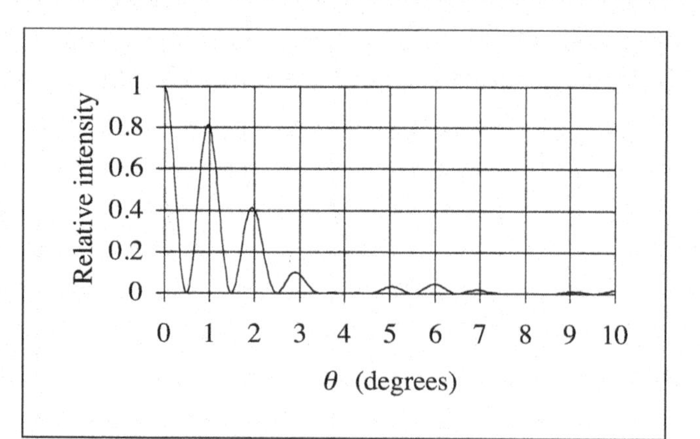

Sketch the relative intensity graph that would result when each of the following changes is made. Your graphs should show the approximate locations of the maxima and minima and be qualitatively correct as to the relative sizes of the maxima. *Explain your reasoning*.

a. The width of both slits is reduced to $a/2$, while keeping the spacing d fixed.

b. The distance between the centers of the slits is reduced to $d/2$, while keeping the width of the slits (a) fixed.

c. A third slit of width a is added to the right of the existing slits, while keeping the distance between adjacent slits (d) fixed.

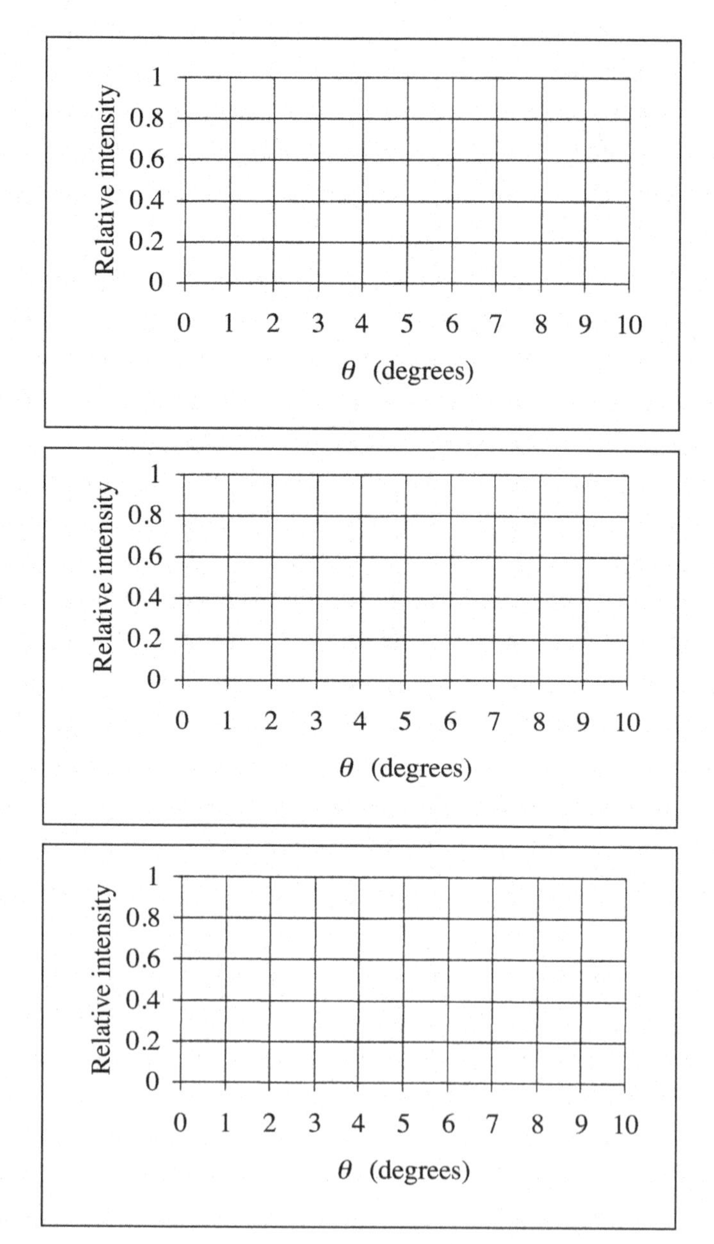

THIN-FILM INTERFERENCE

Name ______________

1. Recall the situation from tutorial, in which light of frequency $f = 7.5 \times 10^{14}$ Hz is incident from the left side of a thin soap film ($n = 4/3$).

   The cross-sectional side view diagram at right extends the ray diagram that you drew in tutorial. It shows the incident ray and several sets of rays resulting from transmission and reflection at the first (left) and second (right) boundaries. The ray that results from the second reflection from the second (right) boundary is not shown.

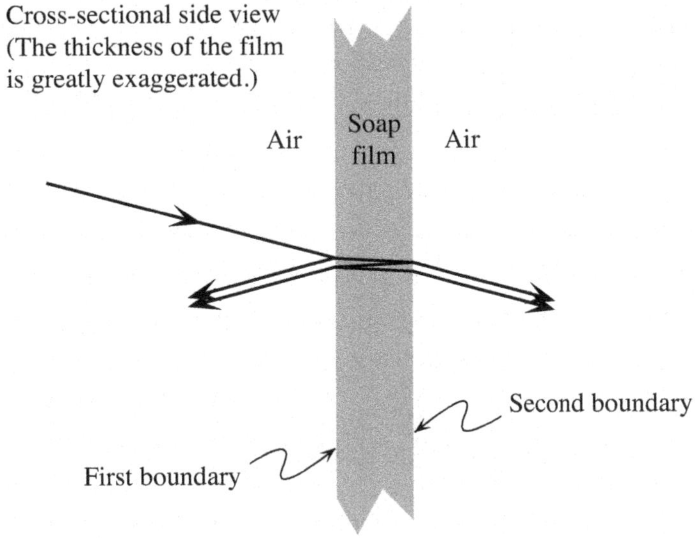

Cross-sectional side view
(The thickness of the film is greatly exaggerated.)

Air | Soap film | Air

First boundary

Second boundary

For the remainder of this problem, assume that the light is incident from the left side of the film at essentially *normal* incidence.

a. For the two rays that pass through the second boundary and into the air, what are the three smallest film thicknesses for which these rays would:

   i. *constructively* interfere? Explain.

   ii. *destructively* interfere? Explain.

b. Recall from section III of the tutorial the pattern that is observed from the *left* side of the film.

   How, if at all, would the pattern seen from the *right* side of the film be different from the pattern seen from the left side of the film? Use your results from part a above to help justify your answer.

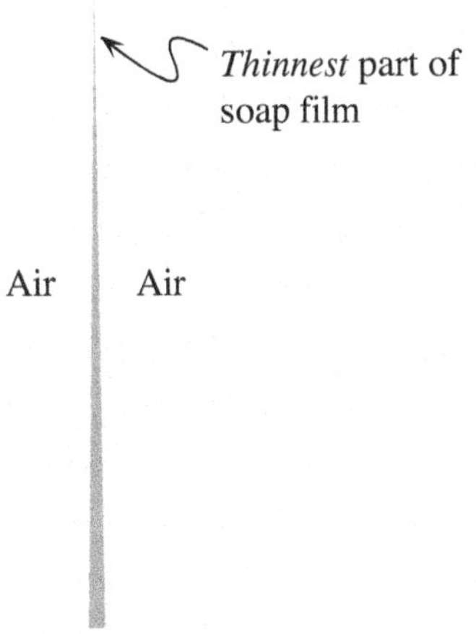

Thinnest part of soap film

Air | Air

Cross-sectional side view
(not to scale)

2. A plate of glass ($n = 1.5$) is placed over a flat plate of plastic ($n = 1.2$). A thin film of water ($n = 1.33$) is trapped between the plates as illustrated in the diagram at right.

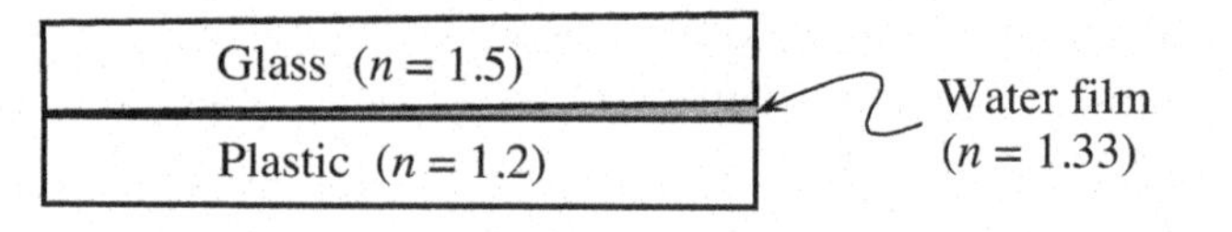

Light of frequency $f = 4.8 \times 10^{14}$ Hz is incident from above at essentially normal incidence. The locations of the dark and bright fringes, as viewed from above, are indicated in the top view diagram at right.

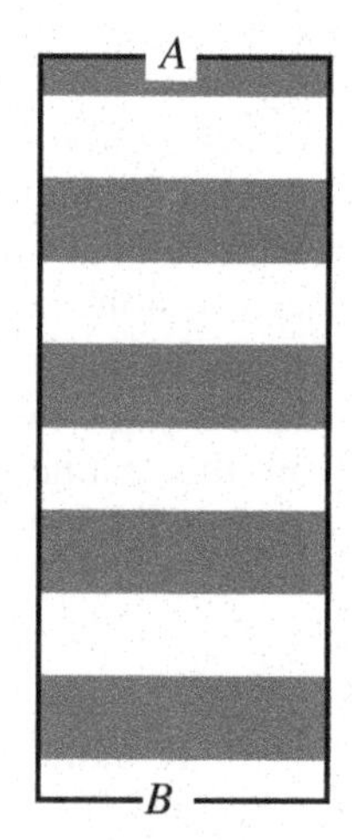

In this problem, ignore reflections from the top surface of the glass plate and the bottom surface of the plastic.

a. Are the plates in contact at point A (where the plate appears dark) or at point B (where the plate appears bright)? Explain your reasoning.

b. Determine the thickness of the gap between the plates at the end where the plates are *not* in contact. (Express your answer in mm.) Show all work.

c. Suppose that the film of water between the plates were replaced with a film of air.

i. Would a bright or dark fringe be seen at the point where the two plates are in contact? Explain your reasoning.

ii. Would adjacent bright fringes be *closer together, farther apart,* or *in the same locations as before?* Explain your reasoning.

1. Identical beams of light are incident on three different pairs of (ideal) polarizers. The double arrow drawn on each polarizer represents its direction of polarization.

   a. Suppose that the incident light in each case were *unpolarized*.

      Rank the three cases (A–C) according to the intensity of the light transmitted past the second polarizer, from largest to smallest. If for any case *no* light is transmitted past the second polarizer, state that explicitly. *Explain your reasoning.*

Case A
45°
Unpolarized light

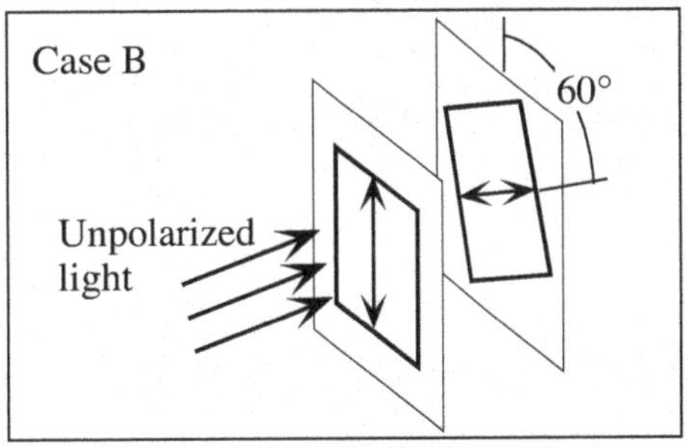

Case B
60°
Unpolarized light

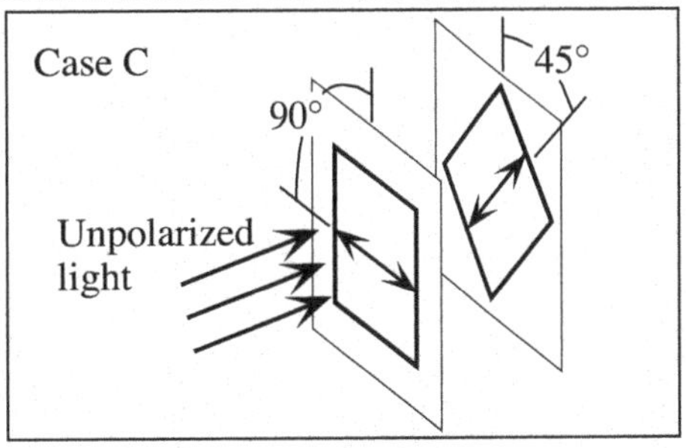

Case C
90°
45°
Unpolarized light

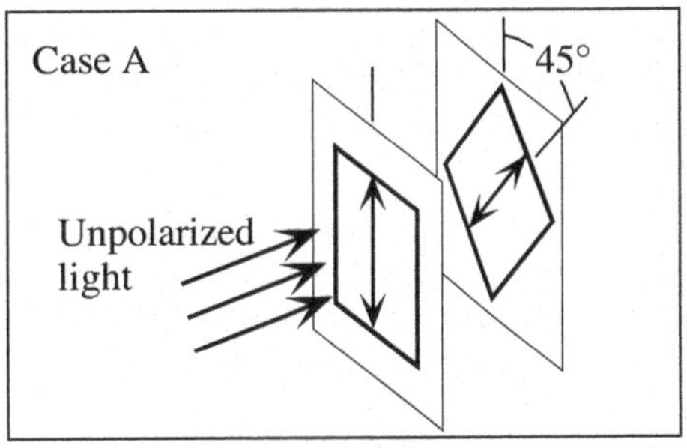

   b. Now suppose that the incident light in each case were *polarized in the vertical direction.*

      Rank the three cases (A–C) according to the intensity of the light transmitted past the second polarizer, from largest to smallest. If for any case *no* light is transmitted past the second polarizer, state that explicitly. *Explain your reasoning.*

2. Unpolarized light of intensity I_o is incident on a pair of (ideal) polarizers, as shown below. The direction of polarization of the first polarizer is vertical. The second polarizer has *unknown* orientation.

For each part below, determine whether it is possible for the light that reaches the observer to have the given value of intensity.

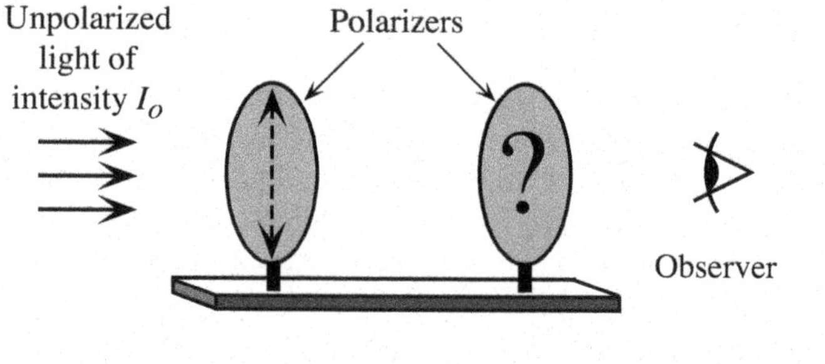

If so: Determine the direction of polarization of the second polarizer. Show all work.

If not: Explain why not.

a. No light reaches the observer.

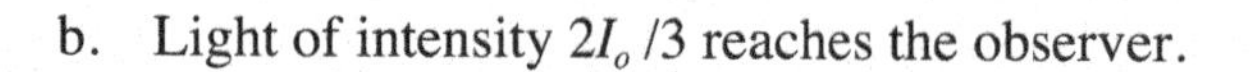
b. Light of intensity $2I_o/3$ reaches the observer.

c. Light of intensity $I_o/2$ reaches the observer.

d. Light of intensity $I_o/4$ reaches the observer.

3. Unpolarized red light is incident on two identical, narrow vertical slits. The photograph at right shows the interference pattern that appears on a distant screen.

Double-slit pattern produced by *unpolarized* red light

a. Specify the quantity or quantities that are adding to zero at the interference minima. ("The light waves from the two slits are adding to zero" is *not* a sufficient answer.)

b. A polarizer is placed directly in front of both slits, so that the light is *vertically* polarized before passing through the slits. It is observed that the intensity at each point on the screen decreases by a factor of two.

How can you account for the decrease in intensity at the interference maxima?

c. The polarizer is slowly rotated through 360°. As the polarizer is rotated, it is observed that the interference pattern does not change.

Consider the following *incorrect* prediction:

> "If the direction of polarization of the polarizer were horizontal, I don't think any light would reach the screen. In that case the light that reaches each slit would be polarized horizontally, so the light would be blocked by each of the vertical slits."

What is the flaw in the reasoning? Explain.

d. Now imagine that one polarizer is placed in front of each slit: one polarizer with its direction of polarization vertical; the other, horizontal.

Would there still be locations on the screen at which the intensity is *zero?* Explain why or why not.

Selected topics

PRESSURE IN A LIQUID

Name ______________________

1. A U-tube filled with water is closed on one end. The tube is about one meter tall. When water is removed from the open end, the water level in the closed end does not change.

   a. What is the pressure at point F before any water is removed?

   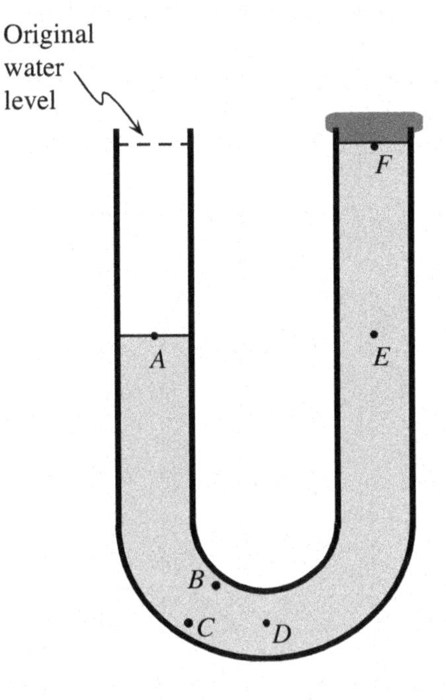

   On the basis of your answer, is there a force exerted by the stopper on the top surface of the water?

   b. The water level on the left is lowered until it reaches point A.

   Does the pressure at point A *increase, decrease,* or *remain the same?* Explain.

   On the basis of your answer, does the pressure at point D *increase, decrease,* or *remain the same?* Explain. If the pressure changes, how does the change in pressure at point A compare to the change in pressure at point D?

   Do the pressures at points E and F *increase, decrease,* or *remain the same?* How do the changes in pressure at these points compare to the change in pressure at point A?

   Does the force exerted by the stopper on the top surface of the water *increase, decrease,* or *remain the same?* Explain.

   c. Suppose that point F is 0.5 m above point E. Determine the pressure at point F. (*Hint:* What is the pressure at point E?) The density of water is $\rho = 1000\ \text{kg/m}^3$, $g \approx 10\ \text{m/s}^2$, and atmospheric pressure $P_o = 1.01 \times 10^5\ \text{N/m}^2$.

   d. Suppose instead that the tube is much taller than 1.0 m. Calculate the distance above point E at which the pressure in the water would be zero (*i.e.*, find the height of water above point E).

   e. Use your answers above and the definition of pressure to explain why the water level on the right remains at point F for a U-tube that is 1.0 m tall.

2. A W-shaped piece of glassware is partially filled with water as shown. Point X is at the same height as the water level in the center of the tube.

   For each of the following points, state whether the pressure is *greater than, less than,* or *equal to* atmospheric pressure. Explain your reasoning.

   - point W

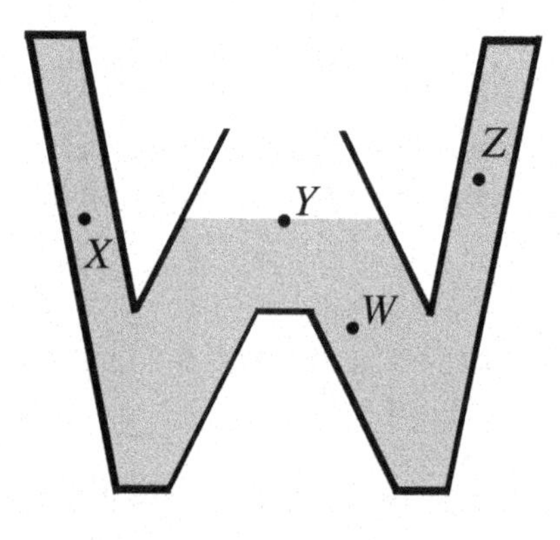

   - point X

   - point Y

   - point Z

3. A U-tube is partly filled with water. Oil is then poured on top of the water on one side. The final water levels on both sides are as shown. The surface of the oil is not shown.

   Points A and D are at the same level; points B and C are at the same level.

   A
   D
   Oil
   Water
   B
   C
   Water

   a. Is the pressure at point B *greater than, less than,* or *equal to* the pressure at point C? Explain.

   b. Will the pressure at the top surface of the oil be *greater than, less than,* or *equal to* the pressure at point D? (*Hint:* What is the pressure at point D?)

   c. Based on your answers to parts a and b, will the top surface of the oil be *above, below* or *at the same height as* the top surface of the water? Explain how your answer is consistent with $P = P_o + \rho gh$.

1. Three objects are at rest in three beakers of water as shown.

 a. Compare the mass, volume, and density of the objects to the mass, volume, and density of the displaced water. Explain your reasoning in each case.

Object 1 floats on top	Object 2 floats as shown	Object 3 sinks
Is $m_1 \begin{pmatrix} > \\ < \\ = \end{pmatrix} m_{\text{displaced water}}$? Explain.	Is $m_2 \begin{pmatrix} > \\ < \\ = \end{pmatrix} m_{\text{displaced water}}$? Explain.	Is $m_3 \begin{pmatrix} > \\ < \\ = \end{pmatrix} m_{\text{displaced water}}$? Explain.
Is $V_1 \begin{pmatrix} > \\ < \\ = \end{pmatrix} V_{\text{displaced water}}$? Explain.	Is $V_2 \begin{pmatrix} > \\ < \\ = \end{pmatrix} V_{\text{displaced water}}$? Explain.	Is $V_3 \begin{pmatrix} > \\ < \\ = \end{pmatrix} V_{\text{displaced water}}$? Explain.
Based on your answers above, is $\rho_1 \begin{pmatrix} > \\ < \\ = \end{pmatrix} \rho_{\text{displaced water}}$? Explain.	Based on your answers above, is $\rho_2 \begin{pmatrix} > \\ < \\ = \end{pmatrix} \rho_{\text{displaced water}}$? Explain.	Based on your answers above, is $\rho_3 \begin{pmatrix} > \\ < \\ = \end{pmatrix} \rho_{\text{displaced water}}$? Explain.

b. Object 2 is released in the center of a beaker full of oil, which is slightly less dense than water. In the space provided, sketch the final position of the block. Explain your reasoning.

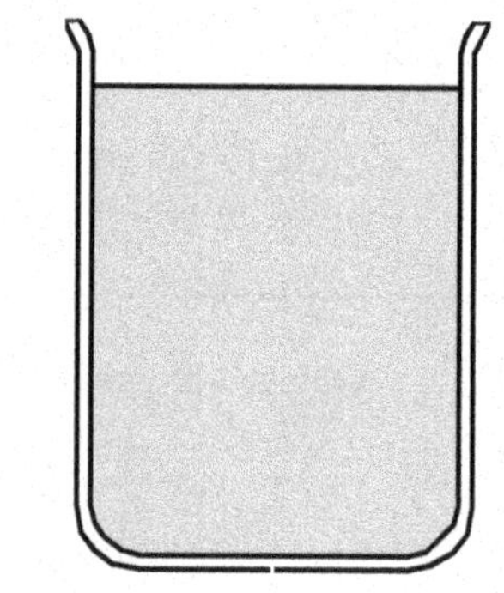

c. On the basis of your answers above, what must be true in order for an object to remain at rest when released in the center of an incompressible liquid?

d. Generalize your answers above to answer the following questions. How does the density of a fluid compare to that of (1) an object that floats in the liquid and (2) an object that sinks in the liquid? Explain.

2. A solid sphere of mass m floats in a beaker of water as shown. A second sphere of the same material but of mass $2m$ is placed in a second beaker of water. In the space provided, sketch the final position of the second sphere.

    a. In its final position, how does the buoyant force on the larger sphere compare to its weight?

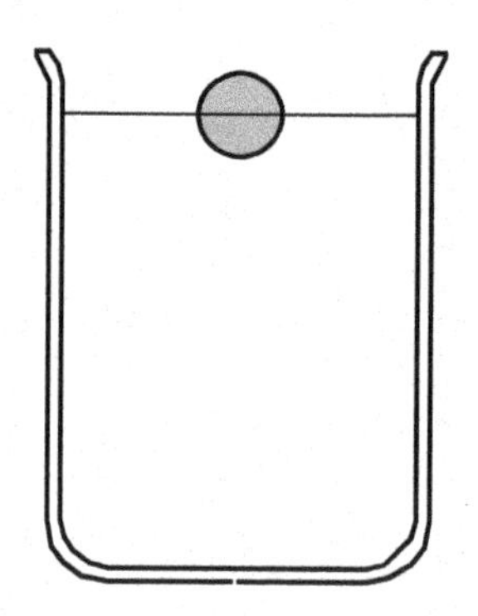

Sphere of mass m

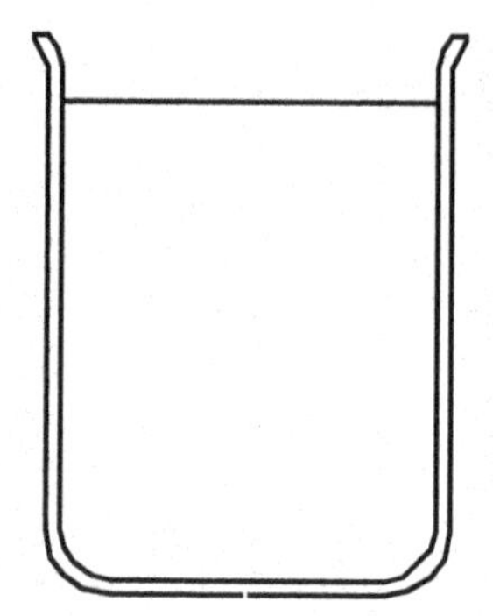

Sphere of mass $2m$

b. How does the volume displaced by the larger sphere compare to that displaced by the smaller sphere?

c. Are your answers to questions a and b consistent with Archimedes' principle? Explain.

3. Two objects of the same mass and volume but different shape are suspended from strings in a tank of water as shown.

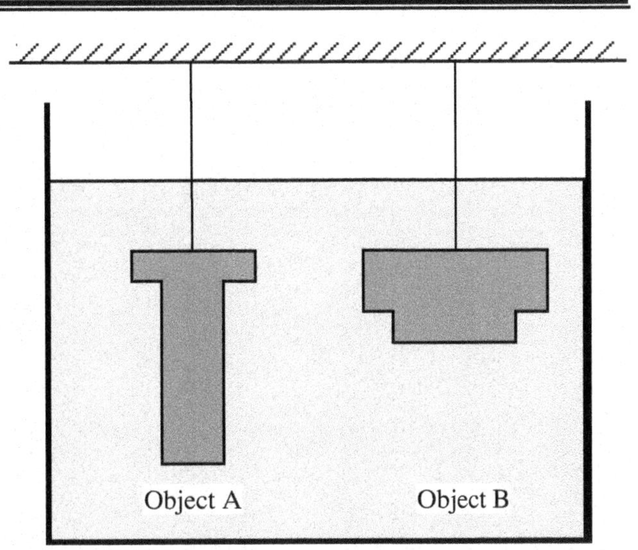

Consider the following student discussion:

Student 1: *"Both objects have the same volume, so both have the same buoyant force. Therefore the tensions in the two strings must be the same."*

Student 2: *"No, that can't be true. The bottom of object A is deeper in the water where the pressure is higher. Therefore the buoyant force on object A must be greater and the tension in that string must be less."*

Student 3: *"I mostly agree with you, student 1. The buoyant force is the same on both objects. However, you forgot the force exerted down on the top of the objects by the water above. That force is larger for object B because the top surface has a greater area, so the tension in the string supporting object B must be greater."*

a. Do you agree with student 1? Explain your reasoning. If student 1 is incorrect, modify the statement so it is correct.

b. Do you agree with student 2? Explain your reasoning. If student 2 is incorrect, modify the statement so that it is correct.

c. Do you agree with student 3? Explain your reasoning. If student 3 is incorrect, modify the statement so that it is correct.

1. a. A cylinder with a valve at the bottom is filled with an ideal gas. The valve is now opened and some of the gas escapes slowly. The valve is then closed, after which the piston is observed to be at a lower position. Assume that the system is in thermal equilibrium with the surroundings at all times.

Ideal gas

Is the final pressure of the gas in the cylinder *greater than, less than,* or *equal to* the initial pressure? Explain.

Explain how your answer is consistent with the forces acting on the piston in the initial and final states.

b. In this process, which of the quantities P, V, n, and T are held constant and which are allowed to change?

c. Consider the following *incorrect* student statement.

"In the ideal gas law, $P = nRT/V$, so the pressure is inversely proportional to the volume. If you decrease the volume, the pressure has to go up."

What is the flaw in the student's reasoning?

d. Explain why it is not possible to use the ideal gas law to determine whether the pressure changed in this process.

2. A long pin is used to hold the piston in place as shown in the diagram. The cylinder is then placed into boiling water.

Pin

Ideal gas

a. Does the temperature of the gas *increase, decrease,* or *remain the same?*

b. Sketch this process in the *PV* diagram at right.

c. Explain why for this particular situation, it is not possible to determine the pressure of the gas as you did on page 1 of the tutorial (*i.e.*, by considering a free-body diagram of the piston).

P

V

1. For each of the following parts, state whether there exists an ideal gas process that satisfies the conditions given. If so, describe the process and give an example from tutorial if possible. If not, explain why such a process does not exist.

   a. There is heat transfer, but the temperature of the gas does not change ($Q \neq 0$, $\Delta T = 0$).

   b. There is no heat transfer, but the temperature of the gas changes ($Q = 0$, $\Delta T \neq 0$).

   c. There is no heat transfer, but work is done on the gas ($Q = 0$, $W \neq 0$).

   d. There is no work done on the gas, but there is heat transfer ($W = 0$, $Q \neq 0$).

2. One mole of an ideal gas is confined to a container with a movable piston. The questions below refer to the processes shown on the *PV* diagram at right. *Process I* is a change from state *X* to state *Y* at constant pressure. *Process II* is a change from state *W* to state *Z* at a different constant pressure.

   a. Rank the temperatures of states *W, X, Y,* and *Z*. If any temperatures are equal, state that explicitly. Explain.

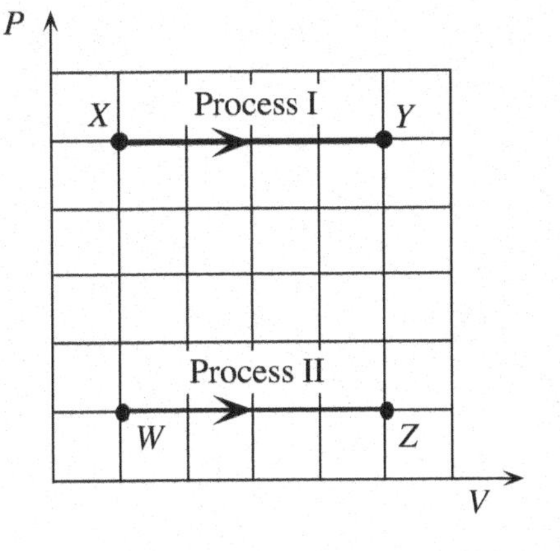

   b. In the two processes, does the piston *move inward, move outward,* or *not move?* Explain.

   c. Based on your answer to part b, state whether the following quantities are *positive, negative,* or *zero*. Explain your reasoning by referring to a force and a displacement.

      i. the work done on the gas during Process I (W_I)

      ii. the work done on the gas during Process II (W_{II})

   d. In Process I, is the heat transfer to the gas *positive, negative,* or *zero?* Explain.

3. Process I from part 2 is used in conjunction with a constant-volume process *(Process III)* and an isothermal process *(Process IV)* to form a cyclic process similar to those used in heat engines and refrigerators.

a. How does the displacement of the piston in Process I compare to the displacement of the piston in Process IV? Explain.

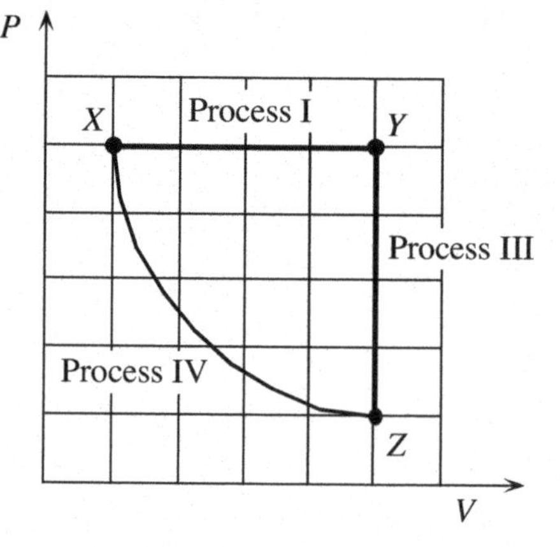

b. Is the *absolute value* of the work done on the gas in Process I *greater than, less than,* or *equal to* the *absolute value* of the work done on the gas in Process IV? Explain your reasoning. *(Hint:* How does the force on the gas by the piston in Process I compare to that in Process IV?)

c. i. How is the work done on the gas during the complete cycle (W_{cycle}) related to W_I, W_{III}, and W_{IV}? Write a mathematical expression.

ii. Is there a region on the graph whose area is equal to the absolute value of the work done on the gas in the cycle? If so, identify that region.

iii. Is the work done on the gas during the complete cycle *positive, negative,* or *zero?* Explain.

d. A student is considering the work done in the cycle:

"The work is given by $P\Delta V$. Since the volume returns to its initial value, the total work done in the process must be zero."

Do you agree with the student? Explain your reasoning.

WAVE PROPERTIES OF MATTER

Name ________________

1. Consider the two-slit electron interference experiment described in the tutorial. (Shown at right is the pattern seen on a phosphorescent screen placed far from the slits.)

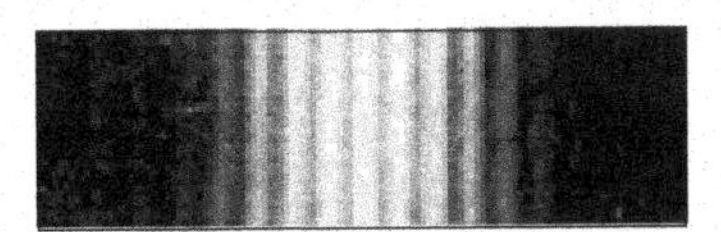

Pattern on phosphorescent screen

Suppose that this experiment were repeated using muons, with each muon having the *same kinetic energy* as each of the original electrons. (Recall $m_\mu \approx 200m_e$.)

a. Is the momentum of each muon *greater than, less than,* or *equal to* the momentum of each of the original electrons? Explain your reasoning.

b. Is the de Broglie wavelength of the muons *greater than, less than,* or *equal to* the de Broglie wavelength of the original electrons? Explain your reasoning.

c. When the electrons are replaced with muons, would the bright regions on the screen be *closer together, farther apart,* or *stay at the same locations as before?* Explain your reasoning.

d. Consider the statement below made by a student:

"Muons have a higher mass than electrons, but because the energy, E, is related to the wavelength by $E = hc/\lambda$, muons that have the same kinetic energy as electrons will also have the same wavelength."

Do you agree or disagree with this statement? Explain your reasoning.

1. Consider the experiment shown in the figure at right. An evacuated tube contains two electrodes, A and B. Monochromatic light of $\lambda = 250$ nm is incident on electrode B, which is made of nickel ($\Phi = 5.2$ eV).

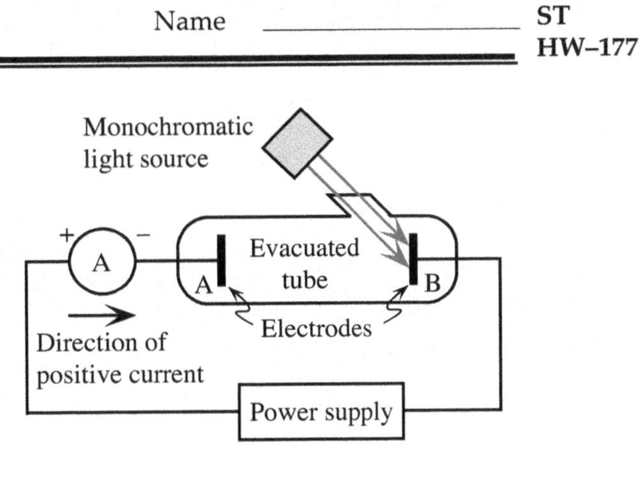

When the voltage across the electrodes is zero volts, the ammeter reads zero current.

Would the ammeter read *zero* current or a *non-zero* current if you were to:

- Double the intensity of the light source? Explain.

- Increase the voltage across the electrodes from 0 V to + 5.5 V? Explain.

- Replace the nickel electrode with one made of aluminum ($\Phi = 4.2$ eV)? Explain.

2. In a photoelectric effect experiment, a certain stopping potential is measured. Suppose that the frequency of light were doubled.

   Would the stopping potential change by a factor of *exactly* 2, *greater than* 2, or *less than* 2? Explain.

Credits:

Page 130: Source: *Physics by Inquiry* by Lillian C. McDermott and the Physics Education Group. © 1996 by John Wiley & Sons, Inc. Reprinted by permission.

Page 133: Source: *Physics by Inquiry* by Lillian C. McDermott and the Physics Education Group. © 1996 by John Wiley & Sons, Inc. Reprinted by permission.

Page 149: Source: *Atlas of Optical Phenomena* by Michel Cagnet, Maurice Françon, and Jean Claude Thrierr. © 1962 by Springer Verlag. Reprinted by permission.

Page 151, 152: Source: *Atlas of Optical Phenomena* by Michel Cagnet, Maurice Françon, and Jean Claude Thrierr. © 1962 by Springer Verlag. Reprinted by permission.

Page 175: Source: "Elektroneninterferenzen an mehreren künstlich hergestellten Feinspalten," C. Jönsson, Zeitschrift für Physik, 161:454, 1961. © 1961 by Springer-Verlag. Reprinted by permission.